The Smart City

How AI is Shaping the Future of Urban Life

by

Emily M. Foster

The Smart City

How AI is Shaping the Future of Urban Life

Contents

Introduction

The trajectory of urban development is on the cusp of a radical transformation. As cities around the globe grapple with unprecedented growth challenges and environmental concerns, the emergence of artificial intelligence (AI) presents both a beacon of hope and a source of intrigue. This book sets out to explore how AI is interweaving itself into the fabric of cities, potentially revolutionizing everything from how citizens commute to the way public safety systems operate. At its core, this is a journey into the heart of contemporary urban life and its future prospects.

Urban areas are not just growing; they're evolving at an incredible pace. The allure of city living offers opportunities for economic advancement, cultural experiences, and social networking, drawing more individuals from rural areas into urban centers. Yet, this massive influx brings about significant challenges—congestion, pollution, and resource scarcity are just a few of the glaring issues. AI, with its capacity for data analysis and predictive reasoning, holds the promise of not just addressing these hurdles but doing so in a manner that propels cities into the future of smart living.

The concept of smart cities is no longer a futuristic fantasy. It's a present-day reality that's taking shape across the world. These cities leverage data and technology to improve services, cut costs, and enhance sustainability. AI stands at the forefront of this transformation, offering innovative solutions that support efficient energy usage, streamlined transportation systems, and improved public safety mechanisms. The potential seems boundless, encouraging urban planners and policymakers to dream bigger and innovate further.

However, the evolution toward smarter urban environments isn't without its concerns. The implementation of AI technology raises questions about data privacy, ethical considerations, and the job market's restructuring. As technologies that once seemed confined to the realm of science fiction integrate into our daily lives, they challenge us to re-evaluate old systems and structures. Ensuring that these technologies work for all citizens and enhance equity, rather than exacerbate existing disparities, is crucial.

An amazing facet of AI's role in urban environments lies in its ability to transform public engagement and governance. By streamlining processes and providing transparent access to information, AI tools empower citizens to participate in civic discussions and decision-making processes more actively. Moreover, smart systems equipped with AI can enhance transparency in government operations, thereby fostering greater trust between citizens and their leaders.

Of course, the effectiveness of AI in transforming cities is tied to the collaboration between various stakeholders. Private companies, public sectors, academic institutions, and citizens must all come together to navigate the complexities of this technological integration. The role of international cooperation and funding also comes into play as cities look beyond their borders for inspiration and support in AI deployment.

As we explore the intricacies of AI-driven smart cities, it becomes apparent that this transformation is not merely a technological shift but a sociocultural one too. It's an ambitious vision that seeks to harmonize urban living with ecological sustainability and social equity, ensuring that cities are not only technologically advanced but also livable and inclusive for all.

The content in this book reflects a comprehensive take on the trajectory of AI and its potential impact on urbanization. While there are immense opportunities on the horizon, the challenges can't be overlooked. Our aim is to create a roadmap for understanding, engaging with, and ultimately thriving in the new urban landscapes powered by AI.

As we delve into this transformative journey, readers are encouraged to envision the possibilities that cities—and their inhabitants—can achieve by harnessing AI responsibly. From paving the way for more efficient public services to protecting the environment, the scope of AI in modeling future cities is a testament to human innovation and adaptability.

In essence, this book is a call to action—a realization that the future of urban life as we know it is changing, and we all have a role to play in shaping its outcome. Whether you're a tech enthusiast, city planner, policy architect, or a curious citizen, there's something here for everyone. Let's proceed with the knowledge that the possibilities are numerous and the journey, though complex, is one worth undertaking.

Chapter 1: The Evolution of Urban Landscapes

Urban landscapes have been continuously evolving, shaped by the dynamic interplay of technological advancements and societal needs. Over the past century, cities have transformed from industrial hubs into sprawling metropolises, each striving to balance growth with sustainability. As we venture deeper into the 21st century, the shift towards smart cities marks a pivotal point in urban development. Smart cities harness emerging technologies like AI to address pressing challenges such as congestion, pollution, and inefficient resource management, all while enhancing the quality of life for their residents. This transformation is driven not just by the potential of AI, but by the collaborative efforts of urban planners, policymakers, and communities to create environments that are more adaptive, responsive, and sustainable. The evolution of urban landscapes is a testament to human ingenuity and the ceaseless pursuit of a better future, where cities not only meet the demands of the present but anticipate the needs of tomorrow.

The Rise of Smart Cities

If we go back just a few decades, the concept of smart cities was more science fiction than reality. Today, however, these cities are reshaping urban landscapes around the world. Smart cities leverage cutting-edge technologies to enhance various facets of urban life, from transportation systems to public safety, aiming for an efficient, sustainable, and livable environment. But what exactly defines a smart city? It's more than just adopting the latest tech—it's a strategic integration where technology, governance, and society work in concert, redefining how cities function and thrive.

At the heart of any smart city lies data. It's the lifeblood that powers the myriad of technological solutions deployed across urban systems. Sensors and IoT devices dot the cityscape, collecting vast amounts of information every second. From air quality meters checking pollutants to motion sensors monitoring traffic flow, data feeds into centralized systems that analyze and optimize urban operations. This connectivity facilitates a responsiveness that's unprecedented, allowing cities to anticipate problems and react swiftly—often in real-time. Rolling out these technologies, though, comes with its share of challenges, such as ensuring data privacy and securing the infrastructure from cyber threats.

The infrastructure of smart cities is inherently designed to be intelligent and adaptive. Consider public transportation networks. These systems are becoming increasingly interconnected, allowing for dynamic route adjustments based on real-time traffic data. This adaptability not only ensures smoother commutes but also optimizes energy usage and reduces emissions. Imagine an AI-driven control center orchestrating city buses, subways, and bikes, seamlessly weaving through a tapestry of urban life. It's not just about getting from Point A to Point B anymore—it's an integrative experience aimed at minimizing friction and maximizing efficiency.

Housing and energy are two fundamental aspects significantly impacted by the rise of smart cities. Innovative building designs incorporate smart tech to control lighting, heating, and ventilation systems, adapting to both the environment and the occupants' needs. Such integration results in substantial energy savings and reduced carbon footprints, tuning the once-static walls and ceilings into dynamic, feel-good spaces. Meanwhile, smart grids play a pivotal role, managing the distribution of electricity with an unprecedented level of precision. They facilitate the integration of renewable energy sources such as solar and wind power, making the urban environment not only self-sustaining but also future-proof.

Public safety and security have always been top priorities for urban areas, with smart cities no exception. The advent of AI-enhanced surveillance and predictive policing offers tools that can help foresee potential threats and deploy resources more efficiently. While this sparks a vigorous debate concerning privacy and ethics, the potential benefits in reducing response times and preventing crime are significant. We find ourselves at a crossroads, balancing the immense potential of AI with the need to protect individual rights and freedoms—an ongoing conversation crucial in shaping the cities of tomorrow.

Moreover, smart cities are reshaping the face of governance. Digital platforms enable a level of transparency and citizen engagement that simply wasn't possible in the analog age. Residents can connect with city officials, voice opinions, and even vote on community issues with the convenience of a smartphone app. It's about empowering citizens and ensuring that their voices are integral to the urban planning process. While technology offers the tools, it's up to the stakeholders to maintain the human touch and moral compass that guide these transformations.

A natural spillover from the evolution of smart cities is the transformation of utility services. Waste management, water distribution, and environmental monitoring systems are becoming smarter, more efficient, and less intrusive. For instance, smart sensors in waste bins signal when they need emptying, optimizing collection routes and reducing fuel usage. Similarly, advanced water management systems predict usage patterns and detect leaks, conserving precious resources in the process. These innovations underscore a central tenet of smart cities: less waste, more efficiency.

The development of these integrated urban ecosystems isn't without hurdles. While the potential for a connected urban future is enticing, it demands substantial investment and cooperation among diverse stakeholders. Policymakers, private corporations, technology developers, and citizens must work together to ensure balanced growth. There's a need for frameworks that protect against data misuse, ensure equitable access to technology, and sustain an inclusive growth model that doesn't leave any segment of society behind.

Despite these challenges, the momentum toward smart cities is undeniable. The rise of such cities represents a seismic shift in how we perceive urban spaces, tranquility, and dynamism co-existing in harmony. As technology evolves, so too will the capacity for our cities to address emerging challenges like climate change, housing shortages, and aging populations. Visionary city planners are increasingly reliant on AI to navigate these intricate issues, drawing inspiration and practical solutions from complex algorithms.

The concept of smart cities isn't static; it's an ever-evolving idea shaped by the interplay of innovation and necessity. It holds the potential to redefine urban life as we know it, making cities more connected, more sustainable, and more resilient. For tech enthusiasts, urban planners, policymakers, and city dwellers, there's no better time to engage with this transformative movement and help sculpt the landscapes of the future.

Urban Challenges in the 21st Century

As cities continue to grow and evolve, they face a myriad of challenges that test both their resilience and adaptability. Urban environments are complex entities, where the convergence of technology, society, and environment results in a dynamic yet often unpredictable landscape.

It's increasingly evident that cities need to address issues spanning from infrastructure strain to environmental concerns. At the heart of these challenges lies a critical question: how can modern cities leverage technological advances, particularly artificial intelligence (AI), to overcome these hurdles and create a sustainable future?

To begin with, urbanization is a relentless force. The United Nations estimates that by 2050, 68% of the world's population will live in urban areas. This rapid urbanization presents both opportunities and challenges. On one hand, cities can be hubs of innovation and economic growth; on the other, they are under immense pressure to provide adequate housing, transportation, and public services. This brings about the challenge of maintaining infrastructure that can keep up with such growth, especially when much of it in older cities is already outdated. Imagine trying to accommodate modern technology in buildings constructed long before the digital age—it's a daunting task.

Moreover, cities are grappling with environmental challenges. From air pollution to waste management, urban areas are often at the center of environmental degradation. Traffic congestion, for example, doesn't just waste time; it contributes significantly to air pollution and greenhouse gas emissions. Urban waste management is another pressing issue, with landfills nearing capacity and the need for effective recycling programs becoming ever more urgent. Furthermore, climate change intensifies these issues, as cities must brace for extreme weather events, which can overwhelm existing infrastructure. In this context, how cities manage their environmental footprint is crucial to ensuring a livable future for their residents.

Housing affordability represents yet another significant challenge. The rise in urban populations often outpaces the construction of affordable housing, leading to increased rent and property prices, which in turn exacerbates social inequality. Many cities find themselves dealing with the gentrification effect, where long-standing communities face displacement due to economic pressures. Ensuring that cities remain inclusive and equitable as they grow is a tricky balance to strike.

Public safety is another area where cities face substantial challenges, complicated by the threats of terrorism, crime, and cybercrime. The integration of AI in public safety can offer innovative solutions like predictive policing and behavior analysis but raises ethical questions around surveillance and privacy. Nonetheless, in their quest to enhance safety, cities are exploring how AI can predict potential threats and improve response times, fostering safer urban environments.

Transportation systems are at breaking point in many of the world's largest cities. Traditional public transportation solutions are often cumbersome and underfunded, making it difficult to address the needs of growing populations. Urban mobility requires a revolution—here, technologies such as autonomous vehicles, ride-sharing platforms, and AI-driven traffic systems present promising prospects. However, the integration of these technologies presents hurdles in terms of public acceptance, regulatory environments, and existing infrastructure adaptation.

Healthcare in urban areas also presents challenges, particularly in ensuring equitable access to services. As cities expand, the demand for healthcare resources increases, placing enormous pressure on existing systems. The integration of AI into healthcare services could offer significant improvements in diagnostics, treatment planning, and patient care. Yet, this must be approached carefully, ensuring that technological advancements don't further widen the gap between different socioeconomic groups.

Digital equity is another major hurdle yet to be fully addressed. The digital divide—disparities between those with easy access to digital technologies and those without—can stifle the growth of smart cities. For AI technologies to be fully realized in urban planning and citizen engagement, it's essential that all citizens have access to the internet and the digitally-enabled services it provides. Bridging this divide is about more than just technology; it's about incorporating digital skills into education systems and community programs.

As urban areas adapt to these challenges, they must also contend with the need for effective governance and decision-making processes. The complexity of modern cities requires agile and forward-thinking governance that can engage diverse stakeholders—from governmental agencies to citizens and private sectors. AI can potentially aid in these governance processes through data analytics, offering insights that can improve decision-making. However, it requires a careful balance, ensuring that AI-powered solutions remain transparent and are used ethically.

Cities of the 21st century are puzzles with many moving parts. Innovators in urban planning and technology must work hand-in-hand with policymakers and communities to not only tackle these challenges effectively but also ensure they are building resilient and adaptable urban landscapes. It's about reimagining what cities can be, using technology as a tool, not a crutch, to develop sustainable solutions that are both practical and visionary. Embracing this evolution means recognizing the challenges and opportunities that come with the territory, and viewing them as catalysts for transformation.

Chapter 2: Understanding Artificial Intelligence

Artificial Intelligence is more than just a buzzword; it's a transformative force that's redefining how cities function and evolve. In urban settings, AI serves as a dynamic tool for tackling complex challenges like traffic congestion, energy consumption, and public safety, while simultaneously unlocking innovative opportunities for sustainable growth. Understanding AI begins with recognizing its core capabilities—such as machine learning, natural language processing, and computer vision—which collectively drive smart urban solutions. By automating tasks, predicting outcomes, and optimizing resources, AI enhances the efficiency of urban systems and services. Yet, as cities integrate AI into their fabric, they must also navigate ethical concerns and ensure equitable access to its benefits. As planners, policymakers, and citizens explore the potential of AI, one thing is clear: embracing this technology responsibly can lead to smarter, more resilient, and inclusive urban environments, paving the way for a future where technology serves humanity's best interests.

Basics of AI in Urban Contexts

Emerging from the vibrant intersection of technology and urban living, artificial intelligence (AI) is becoming an ever-present force shaping cities worldwide. Understanding AI in urban contexts involves recognizing its transformative capacity to improve the quality of life for city dwellers, streamline city operations, and enhance service delivery. Cities, traditionally seen as engines of economic and cultural dynamism, are now morphing into smart urban ecosystems powered by AI technologies.

Fundamentally, AI refers to the capability of machines to perform tasks that typically require human intelligence. In the context of cities, AI can sift through vast amounts of data to provide insights and instant solutions to complex problems. Imagine a city where traffic lights adjust autonomously to reduce congestion or where public services are tailored to individual needs without overwhelming the existing infrastructure. This is not a far-off vision, but a tangible reality being implemented in cities worldwide.

AI's potential in urban environments extends far beyond efficiency and responsiveness. It paves the way for profound shifts in how urban life is experienced and managed. At its core, AI can analyze patterns and generate insights from real-time data. This capability allows cities to anticipate problems before they escalate, manage resources more effectively, and create a seamlessly coordinated urban experience.

One fundamental aspect of AI utilization in cities is data generation and interpretation. Urban areas constantly produce enormous amounts of data, from the movement of public transportation to energy consumption patterns. AI algorithms can process this data at unprecedented speeds and scales, providing city officials with actionable insights. These insights form the groundwork for strategic decision-making, ultimately making urban environments more adaptable and resilient.

Moreover, the integration of AI into urban systems is not just about technological advancement—it's about creating smarter, sustainable cities. With growing urban populations and finite resources, sustainability is essential. AI technologies play a crucial role in this by optimizing energy usage, reducing waste, and managing natural resources more effectively. Cities can significantly cut down on energy consumption by employing AI-driven smart grids, which balance energy supply and demand, thus reducing carbon footprints.

The benefits of AI in urban spaces, however, are not confined to operational improvements. AI is also about personalizing urban services to meet the citizens' unique needs. Imagine healthcare systems where predictive analytics streamline patient care, or local governments that use AI to enhance public safety and reduce crime. The possibilities are as diverse as they are promising, opening up new avenues for personalized citizen engagement.

Yet, while AI brings opportunities, it's also accompanied by challenges that must be addressed vigilantly. The implementation of AI in urban contexts raises pressing ethical considerations. These include issues of data privacy, surveillance, and the potential for algorithmic biases. Ensuring that AI systems are transparent and equitable becomes paramount to secure public trust and support. Policymakers and technologists must collaborate to develop robust ethical guidelines that govern AI's deployment in cities to protect individual freedoms while embracing technological progress.

The role of AI in fostering economic growth and creativity cannot be overstated. By automating routine tasks, AI frees up human capital to focus on complex problem-solving and innovation. It allows urban economies to adapt swiftly to changing dynamics and leverage AI-driven insights to tap into emerging markets. This technological symbiosis fosters a thriving environment for startups and established firms, empowering them to create solutions that make city life more efficient, secure, and enjoyable.

Equally important is preparing the human workforce to thrive alongside evolving technologies. As AI contextually reshapes urban economies, cities must offer educational opportunities that are aligned with future needs. Fostering a culture of continuous learning and adaptation at every level of society will ensure that all citizens can benefit from AI's advantages in the urban sphere.

City planners are increasingly turning to AI to support complex infrastructure projects, simulate urban growth, and improve housing developments. Traditionally, these areas were cumbersome to manage due to their inherent complexity and scale. AI can now model potential outcomes of urban planning decisions, support infrastructure development, and create viable solutions for challenges like affordable housing and sustainable development.

As we delve further into integrating AI into the urban fabric, interdisciplinary collaboration will be essential. Tech developers, urbanists, and policymakers must work together to create AI systems that are harmonious with the communities they'll serve. Such collaboration ensures systems are designed with inclusivity and social equity in mind, paving the path toward cities that are not just smart but also just. The objective is a future where technological prowess complements cultural and social elements, enriching urban life while respecting its heritage.

The journey towards fully AI-integrated urban centers is ongoing, offering unparalleled opportunities while requiring careful reflection and strategic planning. Every implementation must prioritize sustainability, equity, and innovation, aiming for a future where all city residents can thrive in harmonious, resilient urban habitats. Through thoughtful governance and ethical AI practices, urban contexts will undoubtedly unlock the full potential of artificial intelligence, transforming cities into beacons of sustainable, forward-thinking progression.

Key Technologies Driving Innovation

At the heart of artificial intelligence's transformation of urban landscapes are several key technologies fueling an unprecedented wave of innovation. These technologies are not only enhancing efficiency and sustainability in cities but are also paving the way for transformative experiences for urban dwellers. It's crucial to understand the foundational technologies that are making these new possibilities a reality.

One of the most significant drivers of AI innovation is machine learning, a branch of AI that enables systems to learn and improve from experience without being explicitly programmed. Unlike traditional programming methods, machine learning algorithms identify patterns in vast datasets, making predictions or decisions based on new data. This capability is crucial in urban settings, where diverse and large-scale data is abundant, from traffic patterns and energy consumption to public health metrics.

Natural language processing (NLP) is another vital technology that's dramatically reshaping how we interact with the digital world. With NLP, AI systems can understand, interpret, and generate human language, making it easier for residents to engage with city services. From chatbots responding to citizen queries to real-time translation services improving communication in multicultural neighborhoods, NLP is enhancing the accessibility and inclusiveness of urban environments.

Computer vision, a field of AI that enables computers to interpret and make decisions based on visual data, is gaining traction across multiple urban domains. This technology is used in applications ranging from traffic management, where it helps analyze traffic flow and detect violations, to security systems that monitor public spaces for anomalies. Computer vision's ability to process and understand visual information in real time is critical in creating safer and more efficient cities.

The Internet of Things (IoT) plays a crucial role in connecting various urban components, enabling a seamless flow of information.

IoT devices, ranging from smart meters and sensors to home appliances, collect data that AI systems analyze for actionable insights. This interconnected web of devices supports a plethora of city functions, from optimizing energy usage in smart grids to enhancing public safety through intelligent monitoring systems.

Not to be overlooked, edge computing is rising as a significant force in AI technology. By processing data closer to where it's generated, edge computing reduces latency and bandwidth usage, leading to faster response times—an essential requirement for applications like autonomous vehicles and real-time video surveillance. This decentralization empowers local decision-making, making urban systems more responsive and resilient.

The power of predictive analytics is increasingly evident in the way cities plan and operate. This technology leverages historical and real-time data to forecast future trends and events, aiding in everything from disaster preparedness and traffic flow management to anticipating public health needs. Predictive analytics enables city planners and managers to make informed decisions, enhancing the overall efficiency of urban operations.

Moreover, reinforcement learning, a subset of machine learning, is being leveraged to optimize complex urban systems through trial and error. This approach teaches AI systems to make decisions by rewarding them for favorable outcomes. In cities, reinforcement learning is instrumental in optimizing traffic signals, improving energy distribution, and enhancing logistical operations.

Blockchain technology, known primarily for its application in cryptocurrencies, is also emerging as a key driver of urban innovation. Its decentralized and secure nature can ensure transparency and integrity in transactions and data management. Cities can use blockchain to improve processes such as property transactions, identity management, and supply chain transparency, thereby fostering trust and efficiency.

Another game-changing technology is digital twins—virtual replicas of physical entities. By using real-time data and simulations, digital twins help urban planners and policymakers visualize the impact of potential changes in infrastructure and policy before implementing them. This technology enhances the design, monitoring, and optimization of complex urban systems, reducing risks and informing more sustainable development decisions.

Finally, advances in cybersecurity technologies are essential as cities become digitally interconnected. AI-driven security solutions are critical in protecting urban infrastructures from cyber threats, ensuring that the influx of smart devices and systems doesn't compromise safety and privacy. These solutions dynamically adapt to emerging threats, providing a robust defense against the ever-evolving landscape of cyber risks.

In conclusion, the integration of these key technologies is integral to the design and operation of smart cities. By leveraging machine learning, NLP, computer vision, IoT, edge computing, predictive analytics, reinforcement learning, blockchain, digital twins, and AI-enhanced cybersecurity, urban areas can evolve to be more adaptive, efficient, and livable. While challenges remain in deploying these innovations, the potential for creating sustainable and inclusive cities is immense, heralding a promising future for urban environments worldwide.

Chapter 3: Intelligent Traffic Management

In our fast-paced urban landscapes, where every second counts, intelligent traffic management emerges as a beacon of hope for seamless commuting experiences. At its core, AI transforms how cities handle the ebb and flow of vehicles, optimizing traffic lights in real-time, predicting congestion, and rerouting traffic to minimize delays. By analysing vast arrays of data, AI systems forecast traffic patterns and suggest the most efficient routes, reducing fuel consumption and emissions. Cities worldwide are embracing these solutions, with remarkable success stories demonstrating brisk reductions in congestion and improved air quality. Such innovations promise a future where urban transportation is not only faster but also more sustainable, setting the stage for even greater advancements in public infrastructure and quality of life. As we embrace this era of smart traffic systems, the synergy between humanity and machine intelligence is more than a technological feat; it's a gateway to thriving metropolitan ecosystems.

AI in Public Transportation

In the swiftly evolving domain of intelligent traffic management, AI stands out as a transformative force in public transportation. By employing advanced algorithms, machine learning, and data analytics, AI not only boosts the efficiency of public transit systems but also tailors them to better meet the needs of commuters. For urban environments grappling with increased congestion, pollution, and the demand for sustainable solutions, AI offers a multifaceted approach to revolutionizing public transportation.

One of the primary ways AI enhances public transportation is through automated scheduling and real-time tracking systems. AI-powered software can analyze historical data, current traffic conditions, and weather patterns to predict the most efficient routes and schedules. This results in reduced wait times for passengers, optimized fuel usage, and decreased operational costs for transportation authorities. For instance, cities like Singapore and Hong Kong have already implemented AI-driven solutions that adapt bus and train schedules according to real-time demand, which has significantly improved transit service reliability and commuter satisfaction.

Moreover, AI contributes to improving the safety of public transportation. By integrating AI with video surveillance systems, transportation networks can monitor for potential safety risks and respond promptly. These AI systems are adept at identifying unusual patterns or incidents in real time, such as overcrowding in stations or suspicious activities, and can alert the relevant authorities for swift intervention. Such proactive measures not only enhance passenger safety but also ensure a seamless travel experience.

AI's role in public transportation isn't limited to backend operations; it is increasingly noticeable in passenger interactions. Chatbots and virtual assistants, powered by AI, have become valuable assets in providing commuters with real-time information and support. These tools can handle inquiries into route planning, ticket purchases, or delays, rendering human intervention less necessary. This shift allows human staff to focus on more complex tasks and enhances the overall customer service experience.

Committed to sustainability, AI's incorporation in public transportation is also pivotal for environmental conservation. By optimizing routes and improving traffic flow, AI systems minimize unnecessary idling and emissions. Additionally, AI can assist in transition efforts toward electric and hybrid buses, advancing the overarching goal of reducing the transportation sector's carbon footprint. Cities are keen on implementing AI-driven models that prioritize eco-friendly practices, thus forging a path to more sustainable urban living.

Ticket pricing is another domain where AI is making its mark. Dynamic pricing models, facilitated by AI, assess various factors such as time of day, demand, and current service disruptions to offer flexible pricing options. This not only maximizes revenue for service providers but also incentivizes off-peak travel, thus alleviating congestion during peak hours. Such models could significantly alter how passengers perceive and interact with public transit systems, making them more adaptable to varying schedules and needs.

The integration of AI in public transportation extends to infrastructure planning and management. Predictive analytics allow planners to anticipate traffic pattern trends and prepare accordingly. By understanding these patterns, cities can make informed decisions on where to invest in infrastructure enhancements, new routes, or additional service offerings. Planners can utilize AI to simulate scenarios and predict their outcomes, reducing the risk associated with significant infrastructure investments.

Accessibility is another crucial area where AI shows promise. By enabling better route planning and vehicle monitoring, AI helps to ensure that public transit remains inclusive, especially for individuals with disabilities or those dependent on specific transportation provisions. Voice-controlled systems and apps that provide real-time updates or guidance can greatly enhance the mobility and independence of individuals who rely on public transport for their daily commutes.

While AI offers myriad benefits in public transportation, it also presents some challenges. Privacy concerns, for example, arise with the extensive data collection necessary for AI systems to function effectively. It's crucial that cities balance the need for data with individual privacy rights, establishing robust data protection measures and transparency in data usage. Furthermore, the integration of AI technologies demands significant investment and skilled personnel, possibly posing barriers for less affluent cities or nations.

To harness AI's full potential in public transportation, collaboration is essential. Partnerships between technology firms, urban planners, and public transit authorities can help craft innovative solutions tailored to specific urban challenges. By aligning interests and pooling resources, cities can develop infrastructure that substantiates the technical innovations AI brings to public transportation.

In the grand scheme of intelligent traffic management, AI's contributions to public transportation are a paradigm shift. The systems and strategies that AI inspires are not just technical advancements but foundational changes to urban life's routines. The ongoing marriage of AI with public transit offers cities new opportunities to reimagine how transportation can serve their populations more effectively. Through these innovations, urban areas can achieve greater efficiency, sustainability, and accessibility in their transit systems. As AI continues to evolve, so too will its applications, paving the road for ever smarter and greener cities on the horizon.

Reducing Congestion with AI

Reducing traffic congestion, a persistent challenge in urban areas, is becoming increasingly manageable with the advent of AI technologies. By leveraging machine learning algorithms and real-time data, cities can predict traffic patterns and dynamically adjust traffic signals, thereby optimizing the flow of vehicles. These AI systems can quickly respond to unexpected incidents, like accidents or roadwork, rerouting traffic to minimize delays. Additionally, AI enables enhanced communication between different modes of transportation, ensuring smoother transitions and better utilization of existing infrastructure. The promising potential of AI to reduce congestion isn't just about improving commute times—it also contributes to reducing emissions, enhancing road safety, and promoting a more livable urban environment. With continual advancements, AI-driven traffic management is poised to revolutionize how cities confront congestion, paving the way for more efficient and sustainable urban mobility solutions.

Case Studies in Traffic Flow Optimization explores how cities around the world have turned to artificial intelligence to combat congestion, transforming how traffic moves through urban landscapes. By examining specific examples, we'll see how AI is not just an abstract concept but a functional, real-world tool that reshapes daily commutes and urban dynamics.

One prominent example comes from Los Angeles, home to some of the most notorious traffic jams in the United States. In an effort to mitigate gridlock, the city adopted an AI-based adaptive traffic signal system called ATCS. Instead of relying on fixed schedules for lights, the system uses data from road sensors and cameras to adjust signal timings dynamically. This means traffic flow is constantly being optimized based on real-time conditions. Since implementation, Los Angeles has reported a 12% reduction in travel time on average across major corridors, offering a tangible improvement for commuters and a significant environmental benefit through reduced emissions.

Lest this be seen as a uniquely American solution, let's shift focus to Asia, where Singapore has long grappled with the challenges of managing traffic in a densely populated island nation. Singapore's Land Transport Authority employs a sophisticated AI-driven model that includes predictive monitoring and demand forecasting. By analyzing vast datasets derived from monitoring systems across the city, Singapore's AI is capable of not only reacting to current traffic conditions but also anticipating and mitigating future congestion points. As a result, the city-state continues to maintain its reputation for efficient traffic management, despite its population of over five million people.

Over in Europe, the city of Barcelona provides another remarkable case study, showcasing how AI can be integrated into comprehensive urban mobility strategies. Barcelona implemented a "Superblocks" model, reorganizing urban areas to give priority to pedestrians and minimize vehicle traffic in certain zones. AI assists in managing this transition by evaluating traffic patterns and optimizing routes for logistics operations, ensuring that necessary transport can still flow smoothly. Beyond easing congestion, this initiative has greatly improved air quality and enhanced urban livability, creating new spaces for community activities.

In a different context, Pittsburgh, Pennsylvania in the United States showcases the potential of AI to revolutionize vehicle-specific traffic management. The city partnered with a tech company to deploy a smart traffic system using machine learning to optimize bus routes and schedules. Given that buses contribute significantly to traffic volume, this AI-driven approach has led to more predictable transit times and reduced delays without needing vast infrastructural changes. In analyzing millions of data points collected from sensors and GPS systems, Pittsburgh's AI enables a fluid adaptation to evolving traffic conditions, providing a model that other cities with extensive public transport networks could emulate.

Beyond easing congestion, AI's role in traffic flow optimization reflects broader narratives about resilience and adaptability in urban systems. In Mumbai, India, a city notorious for its chaotic traffic, AI systems have been introduced to assist in everything from managing traffic lights to predicting areas likely to become congested. This proactive approach not only alleviates immediate congestion but also enhances city managers' long-term planning by providing actionable insights about evolving traffic patterns. Mumbai's AI program is part of a broader initiative to transform the cityscape into a more manageable, less stressful environment for its residents.

One innovative yet experimental project comes from the Netherlands, where researchers have developed an AI-guided system for managing freeway ramp meters. These AI systems are based on deep learning models that predict traffic flow with high accuracy, ensuring that vehicles merge into traffic streams safely and efficiently. This experiment has shown promise in maintaining smoother speeds on major highways, reducing bottlenecks, and ultimately making highway travel more predictable and pleasant.

The outcomes witnessed from these various projects underscore the potential of AI to learn, adapt, and deliver measurable benefits to traffic systems. Despite varied urban contexts, each of these case studies illustrates a common theme: intelligent traffic systems, powered by AI, are more than just technological feats; they are strategic tools for sustainable urban development.

Yet, embracing AI isn't without its challenges. These systems require significant investments in infrastructure, data management, and operational transformation. There are concerns regarding privacy and the ethical management of vast datasets collected for optimization purposes. Moreover, as reliance on autonomous systems grows, there's a continuing need for robust oversight and adaptability. While the promise of AI in reducing congestion is immense, its long-term success will ultimately depend on how effectively these challenges are addressed. Furthermore, public willingness to embrace emerging technologies plays a pivotal role in the seamless integration of such systems.

In sum, the examined case studies not only highlight the adaptability of AI across various urban settings but also illustrate its potential to transform traffic management from a notoriously inefficient system into one characterized by fluidity and precision. As AI continues to advance, its integration into city planning and management is expected to further expand. Beyond just reducing congestion, AI-driven traffic models may very well lay the groundwork for more comprehensive, adaptive smart city environments that enhance overall quality of life.

The journey towards optimized traffic flows powered by AI is just beginning, and as more cities experiment and implement these systems, the cumulative lessons learned become invaluable. As urban populations continue to rise, the drive towards innovation is not merely aspirational but essential. In this landscape, AI stands as a formidable ally in crafting urban futures where congestion might one day become a relic of the past.

Chapter 4: Sustainable Energy Solutions

The transformation towards sustainable energy solutions is gathering pace as urban environments seek innovative ways to balance growth with ecological responsibility. Emerging technologies like AI stand at the forefront of this transformation, empowering cities to optimize energy consumption, integrate renewable sources, and enhance overall efficiency. By automating complex processes and enabling real-time data analysis, AI supports the intelligent management of energy systems, ensuring that renewable energy not only meets current demands but also anticipates future needs. The development of smart grids is a revolutionary step, allowing for adaptive responsiveness and enhanced energy distribution that conservatively utilizes resources while minimizing wastage. These innovations reflect the growing priority for sustainable urban planning, offering pathways that blend ecological insight with technological acumen. As cities worldwide grapple with increasing energy demands, the push for AI-driven solutions heralds a promising transition towards a greener, more sustainable urban future, demonstrating that with the right tools, cities can become beacons of ecological stewardship and innovation.

AI for Renewable Energy Management

Renewable energy sources, like solar, wind, and hydroelectric power, are cornerstone elements of sustainable city development. However, integrating these diverse energy forms into existing grids requires dynamic management strategies. This is where artificial intelligence steps in, leveraging its predictive capabilities and decision-making prowess to revolutionize the way cities harness and distribute renewable energy.

One of the primary challenges in renewable energy management is the inherent variability associated with natural energy sources. Solar power, for instance, is dependent on weather conditions and time of day, while wind energy can fluctuate with seasonal patterns. AI, through machine learning algorithms, can predict these variations more accurately than traditional methods. By analyzing historical data alongside current meteorological conditions, AI systems anticipate energy production changes, allowing for better planning and utilization of resources.

Smart grids, empowered by AI, are reshaping electricity distribution. At the heart of these systems is the sophisticated capability to balance supply and demand in real-time. During peaks in production, such as a particularly sunny day, AI can direct excess energy where it's needed the most or store it for later use. This intelligent distribution reduces reliance on non-renewable energy backup sources, therefore minimizing carbon footprints.

Furthermore, AI's impact isn't limited to large-scale grid operations. Residential and commercial buildings are benefiting from AI-driven energy management systems that optimize power usage on a granular level. For instance, these systems learn a building's energy consumption patterns and adjust heating, cooling, and lighting accordingly to enhance efficiency. This adaptability not only reduces energy waste but also leads to significant cost savings.

In addition to energy optimization, AI plays a crucial role in maintenance and fault detection. Renewable energy infrastructures, like wind turbines and solar panels, require regular maintenance to ensure efficiency. AI systems equipped with sensor data can predict maintenance needs before issues become critical, minimizing downtime and extending the lifespan of equipment. For example, machine learning models can analyze vibration patterns in turbines and detect anomalies indicative of mechanical stress or wear and tear.

Renewable energy storage remains a critical focus area where AI's role is becoming increasingly prominent. Energy storage technologies, such as batteries, are pivotal, particularly when transitioning to renewable sources. AI optimizes charging cycles and manages discharge rates, extending the lifetime of storage systems while assuring a reliable energy supply even when production is low.

Financially, AI facilitates more attractive investment landscapes for renewable energy projects. By minimizing unpredictability and enhancing efficiency, AI-driven systems make renewable projects more viable and profitable. Investors find reassurance in AI's ability to bolster project returns through predictive analytics and improved operational efficiency.

Beyond economic and operational benefits, AI promotes greater sustainability and environmental stewardship. By reducing energy waste and maximizing renewable utilization, AI helps cities reduce greenhouse gas emissions. This contributes significantly to international environmental goals and commitments, such as the Paris Agreement, which emphasizes the need for smarter, cleaner energy solutions.

As AI continues to evolve, its application in renewable energy management holds promise for even more sophisticated solutions. Innovations such as federated learning edges closer, where decentralized AI systems communicate and learn from each other while maintaining data privacy. Such advancements would enable even broader data sharing and processing power without compromising security, further enriching the global renewable energy landscape.

The integration of AI in renewable energy marks a transformative shift in urban energy management. Smart cities are not just a vision of the future; they are rapidly becoming today's reality, underpinned by intelligent systems that respond and adapt to the intricacies of renewable energy management. This proactive approach ensures a sustainable, resilient urban environment capable of meeting the needs of growing populations while safeguarding the planet for future generations.

In conclusion, AI's role in driving the renewable energy sector forward is undeniably crucial. As we witness the rise of AI and smart technology, it is clear that the potential for optimizing energy systems offers both environmental and economic benefits. The capacity to integrate real-time data, make informed predictions, and manage resources with unprecedented efficiency places AI at the heart of efforts to build sustainable cities equipped to meet the challenges of tomorrow. By investing in AI-driven renewable energy solutions now, cities position themselves not just to survive future challenges, but to thrive as leaders in the global movement toward sustainable urban living.

Smart Grids and Energy Efficiency

Smart grids are revolutionizing the way we manage energy in urban environments by using artificial intelligence to optimize energy distribution and consumption. These advanced systems facilitate the integration of renewable energy sources into the grid, enhancing overall efficiency and reliability. By dynamically adjusting to fluctuations in energy supply and demand, smart grids minimize waste and lower costs, making them a key component in the push towards greener cities. AI-driven analytics assist utility companies and consumers in identifying patterns and opportunities to conserve energy, reduce carbon footprints, and achieve significant cost savings.

These improvements not only bolster sustainability efforts but also empower communities to become more resilient to energy-related challenges, advancing our collective journey towards smarter, more sustainable urban living.

Innovations in Energy Storage are crucial to realizing the full potential of smart grids and enhancing energy efficiency in urban environments. As cities transition toward more sustainable energy systems, the ability to store excess energy produced by renewable sources like wind and solar becomes increasingly important. Traditional energy grids weren't designed to accommodate this variability, causing a need for innovative storage solutions that ensure power supply stability. The progress in this field holds the promise of smoothing out the ebbs and flows of energy production, allowing cities to make more efficient and reliable use of their renewable energy resources.

Lithium-ion batteries have long dominated the energy storage market due to their high energy density and longevity. However, as the demand for energy storage grows, their limitations, such as potential safety hazards and resource depletion, are becoming apparent. Researchers are actively exploring alternatives, including solid-state batteries, which promise enhanced safety and more efficient energy retention. Additionally, advancements in battery technology aim to reduce reliance on rare earth materials, making them both more sustainable and less environmentally taxing to produce.

Beyond traditional battery technologies, flow batteries offer an intriguing solution. They use liquid electrolytes stored in large tanks, permitting easy scaling of capacity. This capability is particularly advantageous for large-scale applications, such as grid stabilization and load balancing in urban smart grids. Moreover, flow batteries can last for decades with minimal degradation, offering a cost-effective long-term storage solution. As these technologies mature, they may overcome key financial and technical barriers, furthering their adoption in smart city infrastructures.

Another promising avenue for energy storage is in thermal energy storage systems, which leverage excess electricity to heat or cool materials that can later be used to reclaim this energy. These systems are particularly beneficial in regions with high heating or cooling demands, effectively aligning energy use with renewable production periods. By utilizing surplus energy during low demand and releasing it during peak hours, thermal storage contributes to overall grid resilience and efficiency.

Flywheel energy storage offers a mechanical alternative, providing quick response times that are invaluable in stabilizing power systems experiencing sudden spikes or drops in electricity. Flywheels operate by increasing rotational speed to store energy and slowing down to release it. This kinetic approach allows for rapid discharge and recharge cycles, potentially making it a key player in frequency regulation within smart grids. As the technology improves, especially in terms of friction reduction and material engineering, flywheels could become a more prominent feature of urban energy landscapes.

The integration of artificial intelligence into energy storage management systems represents another groundbreaking innovation. AI algorithms can optimize energy storage and discharge to cut costs and boost efficiency, adapting to real-time changes in energy demand and supply. Machine learning models are being trained to predict energy usage patterns, allowing for preemptive adjustments in energy storage and distribution, which enhances grid reliability and minimizes waste.

Battery recycling and repurposing are also areas of innovative exploration, addressing the end-of-life aspect of energy storage devices. By developing efficient recycling processes and exploring second-life applications for electric vehicle batteries, for example, cities can significantly reduce waste and resource use. Properly designed recycling programs for lithium and other critical materials support circular economic principles, ensuring that energy storage solutions contribute to environmental sustainability.

Hydrogen fuel cells are emerging as a versatile energy storage solution, converting surplus renewable energy into hydrogen, which can either be stored for long periods or used for other applications, such as powering vehicles. This method of storing renewable energy as hydrogen is particularly attractive due to its potential for integration into multiple sectors, including transportation and industry.

Decentralized energy storage, often facilitated through microgrids, is another exciting frontier. It allows urban areas to localize energy production and storage, reducing the constraints of traditional centralized systems. These microgrids, equipped with local storage, can operate independently when necessary, providing energy autonomy during grid failures and reducing reliance on distant power stations. This innovation is pivotal in building resilient urban infrastructures capable of withstanding disruptions.

Finally, the development of smart energy storage solutions contributes significantly to the goal of decarbonizing urban environments. By enabling efficient use of renewable energy and reducing the need for fossil-fuel-based backup systems, innovative energy storage technologies support cities in their pursuit of carbon neutrality. The synergy between improved energy storage capabilities and smart grids signifies a deeper integration of technology into urban ecosystems, resulting in cleaner, more sustainable cities.

Chapter 5:
AI in Public Safety and Security

In the rapidly evolving landscape of urban security, AI stands as a formidable ally, reshaping the dynamics of public safety. By harnessing predictive analytics and real-time data processing, AI systems assist in crime prevention, offering law enforcement agencies tools to project crime hotspots and deploy resources more effectively and efficiently. These technologies not only enhance preventive measures but also revolutionize emergency response systems, dramatically reducing reaction times during crises and improving overall outcomes. Yet, this progress isn't without its concerns; ethical considerations around surveillance and privacy linger, challenging policymakers to strike a balance between innovation and individual freedoms. As cities worldwide embrace these AI-driven transformations, they pave the way for safer urban environments while fostering discussions on their responsible and equitable implementation.

Predictive Policing and Crime Prevention

As cities continue to evolve and embrace the digital age, public safety and security remain at the forefront of urban planning. At the heart of this progression is predictive policing, an innovative practice aimed at preventing crime before it happens. By leveraging artificial intelligence (AI), law enforcement agencies can anticipate criminal activity with unprecedented accuracy, making our cities safer and more secure environments for all.

Predictive policing hinges on the analysis of big data to forecast potential criminal events. AI algorithms sift through a complex web of historical crime data, social media feeds, and even weather patterns to spot trends and anomalies. This process, akin to data mining, helps identify hotspots where crimes are likely to occur, enabling police departments to allocate resources more effectively. This proactive approach contrasts sharply with traditional policing, which is largely reactive, responding to crimes after they occur.

What's particularly groundbreaking is the use of machine learning in predictive policing. Machine learning algorithms can adapt and improve over time, continuously refining their predictions as they process fresh information. This technology empowers law enforcement to move beyond static crime maps, offering dynamic insights that adjust in real time. It's not just about where crimes are likely to happen, but also when and with what modus operandi. This finely tuned forecasting tool has the potential to keep officers several steps ahead of criminals.

The role of AI in predictive policing extends beyond predictions; it aids in solving crimes faster and more efficiently too. Advanced machine learning models can sift through mountains of evidence, such as video footage or firearms data, to unearth patterns or links that might be overlooked by human investigators. For example, facial recognition technologies allow for rapid identification of suspects in surveillance footage, expediting the investigative process significantly.

But it's not just about solving crimes or reducing crime rates. Predictive policing has the potential to foster trust between law enforcement and communities. By accurately assessing and addressing crime risks, police can engage with communities on a more informed basis, fostering dialogues that promote mutual respect and cooperation. In neighborhoods plagued by crime, the ability of predictive policing to preemptively deploy resources can lead to safer environments and a palpable sense of security.

However, while promising, predictive policing isn't without its challenges and criticisms. Chief among these is the issue of bias in AI algorithms. Machine learning models are only as objective as the data they're trained on, and historical crime data often reflects existing biases. This can lead to a self-fulfilling prophecy, where certain communities, particularly minority and low-income populations, are disproportionately targeted. It's crucial to develop AI systems that are transparent and ethical, ensuring they support equitable law enforcement practices.

Addressing these concerns necessitates a multi-faceted approach that includes rigorous data auditing and a commitment to algorithmic transparency. Law enforcement agencies must work alongside data scientists and ethicists to establish clear guidelines for the development and deployment of these technologies. Public accountability is essential in maintaining the integrity of predictive policing practices, ensuring they serve not just the interests of security, but also justice.

Moreover, privacy and civil liberties remain at the center of the debate around AI in law enforcement. The expansive data required for predictive policing can evoke concerns about surveillance and citizens' right to privacy. It's imperative that urban planners and policymakers forge robust regulatory frameworks that balance the drive for safer cities with the protection of individual freedoms. This includes defining clear boundaries on data collection and establishing oversight mechanisms to guard against misuse.

In a rapidly urbanizing world, where cities are under constant pressure to enhance livability and safety, predictive policing represents a significant step forward. Yet, it's a tool that should be wielded with care, its implementation guided by ethical considerations and community needs. As urban centers experiment with and refine these systems, the opportunity to learn from these experiences grows.

The future of predictive policing is bright, with AI promising even greater advancements. Technologies such as natural language processing could be deployed to analyze textual data from various sources, improving the granularity and depth of crime predictions. Moreover, collaborative efforts across jurisdictions could lead to unified platforms that share data and insights, enhancing predictive capabilities on a broader scale.

In conclusion, predictive policing, when executed thoughtfully and ethically, has the potential to revolutionize public safety in urban areas. By wisely integrating AI, cities can not only thwart criminal activity but also nurture environments where citizens feel safe and protected. The path forward calls for innovation, transparency, and unwavering commitment to the principles of justice and equality.

Enhancing Emergency Response Systems

In the realm of public safety and security, artificial intelligence is revolutionizing how cities respond to emergencies. By harnessing the power of AI, emergency response systems can now better foretell the occurrence of incidents, ensuring quicker and more precise mobilization of resources. For instance, AI-driven data analysis enables faster identification of high-risk areas and trends within urban environments, allowing authorities to strategically disperse emergency responders where they are most needed. Moreover, AI-powered communication platforms streamline information exchange among first responders, delivering real-time updates that enhance situational awareness. This not only sharpens their decision-making process but also boosts the efficacy of their interventions. As cities grow more complex, the integration of AI into emergency response is a compelling advancement toward crafting smarter urban landscapes where safety is prioritized and resilience is fortified.

Ethical Considerations in AI Surveillance Technology doesn't just revolutionize; it also complicates. This is especially true for AI surveillance within the realm of public safety and, more candidly, in enhancing emergency response systems. As AI technologies are integrated into systems designed to safeguard public welfare, ethical quandaries inevitably arise, necessitating a careful balance between innovation and individual rights.

AI surveillance offers a powerful tool for enhancing emergency response systems, providing real-time data, predictive analytics, and rapid decision-making capabilities. Through the lens of public safety, this sounds like an unequivocal win. However, the implications of pervasive surveillance require a deeper consideration of ethics, as it touches on privacy, consent, and potential biases embedded within the technology.

One of the primary ethical concerns is privacy. AI surveillance systems often involve the collection and analysis of vast amounts of personal data, from CCTV footage to location data from personal devices. While these systems can dramatically improve the speed and effectiveness of emergency responses, they also pose significant risks to individuals' privacy. The challenge lies in ensuring that this data is collected and used responsibly, adhering to strict guidelines and oversight to prevent misuse.

Moreover, there's a critical need for transparency regarding how data is collected and utilized. Citizens deserve to know when and how their data is being harvested and for what purpose. This transparency fosters trust in the systems designed to protect them and ensures that they are not unfairly targeted or surveilled. Implementing clear policies and engaging in open dialogues with communities can help in achieving transparency, allowing citizens to understand the role of AI in emergency response and its impact on their privacy.

Consent is another important ethical consideration. While individuals may unwittingly consent to being part of a surveillance network when they enter public spaces, explicit consent processes become vital when personal data extends beyond visual or locational data. Establishing robust frameworks that ensure informed consent is fundamental in maintaining ethical integrity. This can include consent mechanisms that are understandable, accessible, and capable of withstanding scrutiny.

Furthermore, there's the issue of bias within AI systems. Biases can be introduced at various stages, from data collection to algorithm development. In emergency response scenarios, where decisions can mean life or death, the presence of bias can lead to unequal treatment and potentially catastrophic outcomes for marginalized communities. Thus, it's essential to develop AI systems that are fair and unbiased, employing diverse datasets and conducting regular audits to identify and rectify any discriminatory patterns that may arise.

Importantly, the deployment of AI in emergency responses must consider the human element. While AI can rapidly analyze data and offer predictive insights, the importance of human oversight cannot be overstated. Human judgment and empathy are crucial, particularly in crisis situations where nuanced understanding and ethical considerations come into play. Systems should be designed to augment human decision-making rather than replace it, ensuring that the technology acts as a support rather than a determinative force.

Ethical design in AI surveillance is not just about mitigating risks but enhancing the positives, such as inclusivity and accountability. Systems must include input from a diverse cross-section of stakeholders, reflecting the breadth of the communities they serve.

Inclusive design helps in crafting policies and technologies that respect and empower individuals rather than infringe their rights.

Accountability mechanisms should also be robust and well-defined, ensuring that operators and developers of AI systems are responsible for the technology's outcomes. This includes setting up channels for citizens to report issues or concerns about surveillance practices and ensuring that there are appropriate responses to these grievances. A clear accountability structure aids in building public confidence in AI-enhanced emergency systems.

As cities increasingly embrace smart technologies, the need for a regulatory framework tailored to AI surveillance becomes ever more apparent. Such a framework should not just prohibit privacy violations but also encourage innovation in a manner that aligns with ethical norms. It involves ongoing collaboration between tech developers, policymakers, and civil society to create guidelines that safeguard human rights while promoting technological advancement.

Ultimately, the ethical considerations in AI surveillance for public safety and emergency response are not merely about avoiding harm but also about proactively creating systems that are equitable, transparent, and capable of fostering trust. The challenge lies in crafting policies and technologies that not only enhance efficiency and safety but also uphold the values of privacy, fairness, and human dignity. In navigating these complexities, cities can truly harness the transformative potential of AI while respecting the rights of the individuals they are designed to protect.

In conclusion, the ethical landscape of AI surveillance is intricate yet navigable with thoughtful consideration and action. As urban environments continue to evolve with technology, so too must our approaches to ethics in AI, ensuring that we build a future that is both safe and just.

Chapter 6: Healthcare Innovation in Smart Cities

As we delve into the realm of healthcare innovation within smart cities, it becomes apparent that AI has the potential to transform medical services in profound ways. Cutting-edge technologies enable quicker diagnosing and more effective treatment plans, revolutionizing patient care while alleviating pressures on overburdened urban health systems. Remote monitoring, facilitated by AI diagnostics, holds promise for delivering healthcare solutions directly to patients' homes, proving especially beneficial for those with limited mobility or access. Moreover, AI enhances the ability to anticipate and respond to health crises swiftly, improving overall public health preparedness. As smart cities continue to evolve, these innovations not only promise better health outcomes but also forge a more equitable and efficient healthcare landscape for urban populations.

AI-Driven Healthcare Solutions

In the vibrant landscape of healthcare innovation, AI-driven solutions are making a profound impact in smart cities. These advancements are bridging the gap between medical practitioners and patients, ensuring timely, efficient, and precise care. As urban populations grow, demand for healthcare services also escalates, stretching the limits of traditional healthcare systems. AI presents a promising way to address these challenges, offering tools that are not only faster but also more scalable.

One of the most significant roles of AI in healthcare revolves around data. Health data is abundant, generated continuously from diverse sources like medical records, wearables, and health apps. AI technologies, especially machine learning algorithms, are adept at processing and analyzing this large volume of data. By identifying patterns and making predictions, AI aids in early disease detection, thereby shifting the focus from treatment to prevention.

AI-powered diagnostic tools are also a game-changer, drastically reducing the time required for accurate diagnosis. For instance, AI systems can analyze radiology images, such as X-rays or MRIs, with precision that often equals or surpasses human performance. This capability is invaluable in urban hospitals, where a high patient intake can lead to lengthy wait times for diagnoses. Faster diagnostics not only improve patient outcomes but also optimize hospital resources.

Beyond diagnostics, AI contributes significantly to personalized medicine. This approach tailors treatment plans to individual genetic, environmental, and lifestyle factors. AI algorithms sift through vast datasets of patient histories and genetic information to recommend bespoke treatment pathways. By doing so, AI supports physicians in making decisions that best align with a patient's unique health profile, thereby enhancing the efficacy of treatments.

AI-driven healthcare solutions also streamline administrative tasks, allowing healthcare professionals to devote more time to patient care. Automated scheduling systems and AI chatbots can manage appointments, answer basic health inquiries, and process insurance documents with minimal human intervention. This reduces the administrative overhead on medical staff and removes bottlenecks in patient flow.

In addition, predictive analytics powered by AI offer substantial benefits in public health management within smart cities. These tools can forecast health trends and potential outbreaks by analyzing vast datasets of public health information. By recognizing these patterns sooner, city officials can deploy resources proactively, such as vaccines or medical personnel, to areas that need them most.

Telemedicine is another domain where AI's impact is markedly visible. As smart cities strive to ensure healthcare access for all, virtual consultations facilitated by AI provide a convenient and efficient solution. AI assists doctors during virtual consultations by quickly retrieving patient history and suggesting possible diagnoses. This capability becomes even more pivotal in underserved or remote urban areas where physical access to medical facilities is limited.

Moreover, AI is contributing to medical research and development. AI algorithms can process such research material faster than any human team, identifying opportunities for new drugs or treatment methods. This expedites the introduction of innovative therapeutic solutions and extends the frontiers of medical science.

Despite these advancements, integrating AI into healthcare within smart cities is not without challenges. There are concerns about data privacy and security, which are paramount when dealing with sensitive health information. Ensuring that AI systems comply with stringent privacy regulations and standards becomes crucial to maintaining public trust.

Ethical considerations also come into play when deploying AI in healthcare. The impartiality of AI decision-making processes and the need for transparency in algorithms are critical components for ethical AI application in medicine. Addressing these issues is essential to guarantee fair and equitable healthcare access across different urban demographics.

There is also the question of AI's impact on the healthcare workforce. While AI can perform many tasks traditionally carried out by humans, it is not a replacement for the human touch that is invaluable in healthcare. The focus should instead be on how AI can augment healthcare workers, offering them tools to improve their work and providing them with the capacity to focus on interactions that require empathy and human judgment.

Looking ahead, the collaboration between AI developers, healthcare institutions, and city planners will be key in optimizing AI-driven healthcare solutions. As technologies evolve, so must the frameworks guiding their integration into smart city infrastructures. This collaboration can lead to systems that are flexible, adaptive, and responsive to both the needs of individuals and the broader urban community.

Overall, AI-driven healthcare solutions are poised to bring transformative improvements to smart cities, making healthcare more efficient, accessible, and resilient. As stakeholders across sectors engage in dialogue and cooperation, the vision of a seamless and integrated smart city healthcare model becomes increasingly achievable. Through these innovations, the promise of a healthier, more sustainable urban future is well within reach.

Remote Monitoring and AI Diagnostics

In smart cities, the convergence of remote monitoring and AI diagnostics is revolutionizing healthcare delivery, offering opportunities to enhance patient care and streamline medical processes. Through advanced AI technologies, medical data is collected continuously and analyzed in real-time, enabling healthcare providers to detect anomalies and make informed decisions swiftly. This proactive approach not only improves patient outcomes by facilitating early intervention but also eases the burden on healthcare systems by reducing the need for in-person consultations. Wearable devices and smart sensors continuously feed data into AI platforms, allowing for precise monitoring of chronic conditions and immediate responses to health changes, fostering a resilient urban healthcare ecosystem. By integrating these technologies, smart cities are making significant strides toward more accessible and efficient healthcare services, aiming to meet the needs of diverse urban populations while laying the groundwork for future innovations. This transformation demonstrates the power of AI to not just innovate but also humanize healthcare, ensuring that advancing technology benefits all city dwellers by promoting wellness and improving the quality of life.

Improving Access to Healthcare Services has become a cornerstone of healthcare innovation in smart cities, driven largely by the integration of remote monitoring and AI diagnostics. These advancements are laying the groundwork for a transformation in how urban populations access healthcare. By leveraging technology, we can bridge the gap between patients and providers, making healthcare more accessible, efficient, and personalized than ever before.

As cities grow denser, the traditional healthcare model struggles to keep up with increasing demand. Vastly populated urban areas exhibit disparities in healthcare access, with many residents having limited or no access to essential services. AI-powered solutions, coupled with remote monitoring technologies, provide a powerful antidote to these challenges, helping ensure that healthcare reaches everyone, regardless of their location or economic status.

Remote monitoring devices are changing the dynamics of patient care. Wearable technology and home-based diagnostic tools enable continuous health tracking, moving some aspects of health management from clinics into the everyday environments of individuals. These devices can track vital signs, detect anomalies, and alert healthcare providers to any potential issues, allowing for timely interventions. This continuous stream of data empowers patients to be more proactive about their health without physically visiting healthcare facilities frequently, thus easing the burden on these institutions.

AI diagnostics and remote monitoring together offer a significant reduction in healthcare costs. In smart cities, the integration of these technologies helps decrease the need for in-person consultations, thus lowering expenses related to outpatient visits and hospital admissions. By addressing health issues in their nascent stages through constant monitoring, the system can prevent conditions from escalating to a point where costly interventions would be necessary. In turn, this makes healthcare more affordable and accessible for individuals who might otherwise shy away from seeking necessary care due to financial constraints.

The ability of AI to analyze vast amounts of data quickly and accurately ensures that healthcare solutions are not only prompt but also highly personalized. Algorithms can discern patterns and predict health outcomes based on an individual's unique data, paving the way for tailored treatment plans that reflect individual needs and circumstances. Personalized medicine, supported by AI, is remaking patient care, enabling interventions that are more precise and effective than those founded on general population data.

Furthermore, AI diagnostics have shown superiority in detecting diseases such as diabetes, cardiovascular conditions, and even various forms of cancer at earlier stages than ever before. Early detection is critical for improving prognoses and survival rates, and AI is equipped to exceed human capabilities in identifying subtle signs that can signal the onset of serious health issues. The combination of AI and remote monitoring could potentially save millions of lives by ensuring that diseases are caught and treated before they advance.

Telemedicine platforms, enhanced with AI and remote monitoring, are another breakthrough that enhances access. Urban dwellers with mobility issues or those who live in remote parts of a city can have virtual consultations with healthcare professionals, supported by real-time data from their monitoring devices. This setup not only improves access but also increases the flexibility of healthcare, accommodating the diverse schedules and lives of city residents.

For policymakers and urban planners, these innovations offer insights on rethinking healthcare infrastructures in smart cities. By prioritizing investments in technology-driven healthcare solutions, cities can alleviate strain on traditional healthcare services while meeting the demands of burgeoning urban populations. Strategies that integrate AI and remote monitoring represent a fundamental shift towards a more sustainable and responsive healthcare ecosystem.

However, the path to these advancements isn't without challenges. The integration of AI and remote healthcare solutions raises concerns surrounding data privacy and ethical considerations, topics that need careful navigation. Ensuring the security of sensitive health data is paramount to maintaining public trust in these technologies. Moreover, as with any new technology, there is a risk of widening the digital divide, making it essential to ensure equitable access to these innovations across all socioeconomic layers of urban society.

Despite these challenges, the potential to reshape healthcare access and delivery within smart cities is immense. The dream of universal healthcare access, long hindered by logistical and economic barriers, is becoming increasingly attainable thanks to AI and remote monitoring. As cities continue to adopt these technologies, the overarching goal will be to ensure that the promise of improved access translates into tangible health benefits for all urban residents, fostering healthier and more equitable communities.

In conclusion, the fusion of remote monitoring and AI diagnostics in the healthcare domain is not merely a technological innovation; it represents a paradigm shift with fundamental implications for how healthcare services are accessed and delivered in smart cities. As we continue the journey of integrating these advanced technologies into urban life, the vision of an inclusive, responsive, and efficient healthcare system is steadily becoming a reality.

Chapter 7: Digital Governance and Citizen Engagement

As cities integrate AI technologies, the realm of governance is undergoing a profound transformation, marked by increased transparency and citizen involvement. Digital governance offers the promise of more efficient public services, tapping into AI's capability to process data rapidly, predict future needs, and streamline decision-making. Citizens can interact with municipal systems in ways previously unimaginable, leveraging AI-driven platforms to voice their concerns and participate actively in the civic process. However, this shift isn't without its hurdles. Ensuring equitable access to digital tools remains a priority to prevent disparities among populations. By embracing AI, communities have a unique opportunity to bridge the digital divide and foster a more inclusive dialogue between governments and the people they serve, encouraging civic engagement and collaboration that can propel urban environments toward a smarter, more sustainable future.

AI for Transparent Governance

Incorporating AI for transparent governance is revolutionizing how cities operate, making government processes more accessible and understandable to citizens. At its core, transparency involves providing the public with insight into governmental operations and decisions.

This isn't just about releasing data but creating an environment where citizens feel informed and engaged with their local governments. AI acts as a catalyst in this transformation, leveraging data analytics and natural language processing to decode complex information into accessible insights.

Understanding the role of AI in streamlining governance requires examining how it dissects vast volumes of data to identify trends and highlight essential information for policymakers. By processing large datasets, AI can help officials make evidence-based decisions, improving both efficiency and accountability. This, in turn, builds trust between citizens and government entities, as decisions appear less arbitrary and are grounded in objective analysis. Moreover, AI is instrumental in predictive analytics, offering the capability to anticipate societal needs and adapt services accordingly, thus avoiding reactive governance models.

Citizen engagement platforms powered by AI are another layer in transparent governance. These platforms utilize AI to enhance communication channels between the government and its citizens. For instance, chatbot interfaces and virtual assistants can answer frequently asked questions, provide updates on public projects, or display budget breakdowns in real-time. Such technologies ensure that response times are quick and information dissemination is consistent and unbiased. The public can voice concerns, participate in digital town halls, and contribute to decision-making processes more effectively, thereby enhancing civic participation.

Additionally, AI-driven dashboards and visualization tools are game-changers in explaining bureaucratic data. Advanced visualization techniques transform raw data into graphical representations that are easy to understand. These tools can display everything from budget allocations to the progress of public infrastructure projects. When information is organized and accessible, it becomes easier for citizens to hold their governments accountable. Publicizing these visualizations fortifies the notion of openness and responsiveness, crucial elements for democratic societies.

There's an inspirational aspect to using AI in governance. It can unlock potential by connecting citizens with projects or causes they care about. AI can personalize information delivery, suggesting local community initiatives or policymaking sessions based on individual interests. This fosters a sense of involvement and ownership, as people are encouraged to contribute to activities that resonate with them personally. Consequently, AI not only aids transparency but also enriches the communal fabric of urban environments.

Of course, the deployment of AI in governance isn't without challenges. Privacy issues often come to the forefront when discussing AI's analytical capabilities, particularly concerning sensitive personal data. Governing bodies must establish robust legal frameworks to handle data ethically and responsibly, ensuring citizens' privacy is safeguarded. Moreover, there needs to be an active dialogue between AI developers, policymakers, and the public to continuously align expectations and address concerns effectively.

AI also faces the issue of accessibility and digital literacy among citizens. While AI offers transparency, it's imperative that all citizens, regardless of their technical background, can access and comprehend the tools provided. Bridging this gap involves offering education and resources that demystify AI technologies for the average citizen, empowering them to fully engage with enhanced governance processes.

Another critical area to consider is the potential for bias in AI algorithms, which can inadvertently perpetuate existing inequalities. It's crucial for governance systems to frequently audit algorithms and datasets to prevent biased outcomes. Transparency itself extends to how AI decisions are made, requiring governments to make decision-making processes clear and accountable. Constant monitoring and adjustments are necessary to ensure equitable treatment and accurate representations within AI systems.

AI for transparent governance isn't a distant future; it's an ongoing evolution that promises to reshape urban landscapes. Its implementation fosters smarter, more inclusive cities where every citizen has a voice. As these technologies mature, cities must embrace flexibility and adaptability, ensuring that AI serves as a force for good—promoting transparency, accountability, and citizen engagement. The journey to a fully transparent governance system isn't simple, but with strategic planning and collaboration among various stakeholders, the possibilities for enhanced urban governance are endless.

Enhancing Public Services with AI

AI is reshaping how cities deliver public services, making them more efficient, accessible, and responsive to community needs. From streamlining administrative tasks to tailoring personalized services, AI-powered tools enhance everything from healthcare delivery to waste management. By analyzing vast amounts of data, AI helps identify patterns and predict future needs, enabling local governments to allocate resources effectively and improve citizen satisfaction. This proactive approach means cities can address issues before they escalate, ensuring smoother operation of services. Furthermore, AI fosters transparency and accountability by providing real-time insights and feedback on service delivery, which empowers citizens to engage more actively with governance processes. Ultimately, the integration of AI into public services isn't just about efficiency; it's a step towards creating more inclusive and participatory urban environments, paving the way for a brighter and more connected future.

Bridging the Digital Divide is one of the most pressing challenges that cities face today, especially as they strive to enhance public services with AI. In an increasingly digital world, access to technology and the internet is not a luxury; it is fundamental for meaningful citizen engagement with digital governance. However, this access is not evenly distributed, leading to a digital divide that can exclude significant segments of the population from the benefits of AI-driven public services.

The digital divide refers to the gap between individuals who have access to modern information and communication technology and those who do not. This divide is often influenced by socioeconomic factors, such as income, education level, geography, age, and even ethnic background. Bridging this gap is crucial because AI-enhanced services promise improved efficiency, transparency, and accessibility in urban governance, but only for those who can access and use them. By diminishing the digital divide, cities can ensure that all citizens benefit from advancements in technology, thus promoting more equitable urban development.

Bridging this divide involves a multifaceted approach. It is not just about providing access to devices and the internet. It also includes digital literacy, the skills and knowledge needed to effectively use the technology available. AI-driven solutions can be complex, and without the proper skills, citizens may struggle to utilize these technologies fully. Therefore, efforts to enhance digital literacy are as critical as improving physical access. Cities can implement community training programs focused on making digital tools more understandable and accessible, leveling the playing field for interaction with digital governance.

The infrastructure itself is essential. Bridging the digital divide requires a commitment to developing and improving technological infrastructure, such as broadband networks, which can often be lacking in economically disadvantaged or geographically remote areas. Public-private partnerships can play a pivotal role in expanding this infrastructure. By investing in robust digital networks, cities can make pivotal strides toward a more connected populace, ensuring that no resident is left behind in the digital age.

Moreover, smart policies are needed to ensure the inclusion of marginalized communities in the digital future. Cities should focus on crafting digital inclusion strategies that address the specific needs of underrepresented groups. This includes affordable pricing policies, subsidizing internet access for low-income families, and creating digital inclusion funds aimed at helping communities earmarked as vulnerable. Inclusivity in digital policy not only transforms governance but also strengthens the social fabric by fostering a sense of belonging and participation.

AI, ironically, can play a role in bridging the very digital divide it creates. With its capability for personalized learning and adaptive technology, AI can tailor educational programs to meet the varied needs of learners. For example, AI-driven platforms can modify how information is delivered based on the user's current competence, making learning more effective. This personalized approach can help improve digital literacy and keep citizens engaged with the digital tools at their disposal.

However, there are inherent challenges and ethical considerations in deploying AI to bridge the digital divide. Privacy concerns, potential biases in AI algorithms, and the need for transparent decision-making processes are all critical factors that need attention. Cities must navigate these concerns carefully to maintain trust between municipal bodies and residents and to ensure that AI-enforced policies enhance rather than inhibit inclusivity.

The societal benefits of successfully closing the digital divide are immense. When barriers to access are removed, citizens are empowered to participate more actively in civic processes, from engaging in local decision-making to accessing digital public services efficiently. This has wider implications for societal equity and economic growth, where the equitable distribution of digital services can lead to more prosperous and harmonious communities.

Ultimately, bridging the digital divide is not just about technology; it's about justice and opportunity. In fostering an environment where digital governance is accessible to everyone, cities can become hubs of innovation and equitable growth. As AI continues to promise transformative changes in public services, the commitment to bridging the digital divide must remain steadfast, ensuring that all citizens can partake in and contribute to the increasingly digital domains of public life.

Chapter 8: Urban Planning and Design

Urban planning and design are embarking on a transformative journey, fueled by the innovative power of AI. In today's rapidly evolving cities, AI is becoming a pivotal tool in crafting resilient and adaptable urban environments. With the capabilities to analyze vast datasets, AI aids planners in making informed decisions regarding city infrastructure, predicting urban growth patterns, and optimizing the layout of green spaces. This technology enables a more precise approach to designing spaces that harmonize residential, commercial, and recreational needs while ensuring sustainability. AI's predictive analytics offer invaluable insights into future urban dynamics, allowing cities to anticipate and respond to population shifts, climate challenges, and economic changes. As urban landscapes grow more complex, AI serves as a beacon, guiding us towards more efficient, livable, and ecologically friendly cityscapes. The challenge lies in balancing innovative design with the nuanced needs of diverse communities, ensuring cities remain vibrant hubs of life and culture.

AI in City Infrastructure Development

In the complex tapestry of urban planning and design, city infrastructure development is pivotal, more so now with the advent of artificial intelligence. The integration of AI into this sphere is a transformative power, reshaping how we envision, construct, and maintain our urban landscapes. Successfully weaving AI into the fabric of city infrastructure offers a myriad of opportunities to enhance not just the efficiency and functionality but also the sustainability and resilience of cities. This section delves into how AI is revolutionizing the core structure of cityscapes, ensuring they can meet the multifaceted demands of modern life while paving the way for future advancement.

AI-driven approaches to infrastructure development start with making sense of the vast quantities of data generated by urban environments. Every city produces immense streams of data daily, from sensors, geographical mapping, public transport systems, and energy grids, among others. Through sophisticated AI algorithms, this data can be harnessed to conduct real-time analyses, providing urban planners and designers with invaluable insights. These insights can be utilized to optimize the design of public spaces, predict future urban needs, and respond proactively to potential infrastructural challenges before they become crises.

Moreover, AI offers significant advancements in predictive maintenance of infrastructure systems. Traditional methods would require regular manual checks and maintenance schedules, often leading to unnecessary expenditure or unexpected failures. AI systems, on the other hand, continuously monitor infrastructure via sensors to identify patterns that precede system breakdowns, essentially predicting when maintenance should be performed. This predictive capability not only reduces costs but also minimizes service disruptions, thus enhancing the reliability and efficiency of urban infrastructure.

Transportation infrastructure is one area that stands to benefit immensely from AI integration. As cities expand, the demand for effective and efficient transport systems grows. AI algorithms can optimize traffic flow, prioritize public transportation routes, and manage the logistics of freight movement within urban areas. For example, AI can be employed to simulate traffic scenarios, assess the impacts of different traffic management policies, and implement adaptive traffic signals that respond dynamically to changing traffic conditions. These developments contribute to reduced congestion, lowered emissions, and generally more livable urban environments.

But the influence of AI in city infrastructure development is not confined to transportation alone. The construction sector is witnessing its own AI revolution. AI technologies are now capable of automating the design process, performing complex calculations needed for structural engineering, and even managing construction sites with robotic equipment. AI systems can offer design recommendations that balance structural integrity with sustainable practices, such as optimizing materials usage to reduce ecological footprints, thereby ensuring that new constructions align with sustainable development goals.

Cities also increasingly rely on AI to enhance water and energy infrastructure. Sophisticated AI applications help in forecasting water demand, detecting leaks in water infrastructure, and managing resource allocation efficiently. Similarly, AI fridges deep into energy grids, facilitating the delivery of renewable energy, optimizing electricity distribution, and vastly improving energy efficiency in urban areas. These AI applications contribute to creating more sustainable urban environments by reducing waste, enhancing resource efficiency, and improving resilience to environmental stresses.

The creation and nurturing of "smart" infrastructure through AI involve stakeholders across the spectrum, from government entities to private companies and research institutions. These collaborations are vital as they foster innovation and facilitate the implementation of AI solutions. Furthermore, AI-backed smart infrastructure is instrumental in fostering social equity. By providing equal access to essential urban services, AI systems can help ensure that modernization initiatives cater to all residents, thus contributing to an inclusive urban development narrative.

Despite the promising opportunities, the integration of AI in city infrastructure development isn't without challenges. Concerns about data privacy and cybersecurity loom large, necessitating robust regulatory frameworks to safeguard citizen data. Additionally, ensuring AI systems are free from biases rooted in the data they are trained on is crucial for developing equitable infrastructure solutions. Overcoming these challenges requires dedicated efforts towards establishing trust in AI technologies through transparency and accountable practices.

Ultimately, the synergy between AI and city infrastructure development holds the potential to redefine urban environments, making them smarter, more sustainable, and more adaptable to changing requirements and unexpected events. The journey toward fully embracing AI in infrastructure is ongoing, requiring continuous dialogue among planners, technologists, policymakers, and citizens to navigate the complexities and harness the full potential of AI in crafting the cities of the future.

As cities continue to evolve, AI's role in refining and advancing urban infrastructure will grow ever more critical. By leveraging these technologies to their fullest potential, we can build urban environments that not only meet the needs of their inhabitants today but are also prepared to adapt and thrive amidst the challenges of tomorrow. This forward-looking approach is essential for fostering cities that are genuinely reflective of technological advancements and the aspirations of their populations.

Predictive Modeling for Urban Growth

In the dynamic realm of urban planning, predictive modeling stands as a beacon, providing insights into how cities can evolve sustainably amid growing challenges. These models harness the power of AI to simulate and forecast urban expansion, offering a glimpse into the future needs of infrastructure and resources. By analyzing diverse data points, from demographic shifts to economic activities, they equip planners with a sophisticated toolkit for crafting adaptable urban environments. Such foresight ensures not only efficient land use but also the careful integration of green spaces and public amenities, thereby enhancing residents' quality of life. As cities worldwide grapple with rapid growth, predictive modeling emerges as an invaluable ally, fostering urban landscapes that are both resilient and responsive to change.

Green Spaces and AI serve as vital lungs for our urban environments, offering a respite from the dense cityscapes while providing ecological and social benefits. Integrating artificial intelligence into the design and maintenance of these spaces represents an exciting frontier in urban planning and design, particularly within the realm of predictive modeling for urban growth. The challenges of balancing rapid urban expansion with the need to maintain and even increase green spaces are complex. Thankfully, AI offers innovative solutions, enabling smarter, more sustainable urban environments.

AI-driven predictive modeling can forecast urban growth patterns, identifying areas where increased population density might overburden the existing green spaces. By analyzing historical data, demographic trends, and urban dynamics, AI models can propose optimal locations for new parks or enhancements to existing spaces. This anticipatory approach helps urban planners create green zones that not only serve current requirements but also cater to future community needs.

Furthermore, AI can offer insights into the best practices for managing these green spaces, from choosing the most sustainable plant species to implementing efficient irrigation systems. By analyzing weather patterns and soil conditions, AI technologies can recommend tailored maintenance plans that reduce water usage and enhance plant health. This leads to more resilient green spaces that can better withstand the pressures of climate change and urban heat islands.

Consider the value of incorporating AI sensors and Internet of Things (IoT) technologies in monitoring environmental conditions within parks and urban forests. These sensors can track everything from air quality to soil moisture, sending real-time data to AI systems that optimize resource allocation. This precision ensures that resources like water and fertilizers are only used when necessary, minimizing waste and promoting ecological balance.

AI also supports biodiversity within urban areas by identifying suitable habitats for different species and predicting the impact of urban development on local ecosystems. Imagine AI systems generating models that illustrate how different types of urban vegetation can support various wildlife populations. Such insights help planners design urban spaces that encourage biodiversity, blending urban life with nature.

Moreover, AI can influence urban design by simulating potential green infrastructure projects' benefits. By incorporating variables like temperature reduction, pollution filtration, and community wellness into their models, AI tools provide a comprehensive view of how new green spaces can transform urban living conditions. This makes it easier for stakeholders to see the tangible benefits of investing in green infrastructure, fostering increased support and collaboration.

Green spaces have the power to promote social well-being, offering areas for recreation, exercise, and relaxation. In this social context, AI can analyze patterns of park usage, helping authorities understand peak times and preferred activities. This data-driven insight enables the design of spaces that align with the community's needs and priorities, ensuring these areas remain vibrant and engaging.

Integrating AI in green space planning also supports public health initiatives. By measuring foot traffic and activity levels, AI can assist in evaluating a community's engagement with outdoor spaces and creating programs that encourage more active lifestyles. This proactive approach addresses public health challenges related to sedentary living, underscoring the broader societal value of AI-driven green space initiatives.

Incorporating AI into urban planning isn't without its challenges. Planners must consider the technology's ethical and social implications, including data privacy and equitable access to green spaces. Balancing these concerns with the benefits AI brings to green space management requires a nuanced approach, emphasizing transparency and inclusivity.

Ultimately, AI's role in enhancing green spaces is just one component of the broader urban planning landscape, but its impact can be profound. As urban areas continue to grow, AI technologies offer a path toward creating more livable, sustainable, and equitable cities. Through innovative design and strategic planning, AI plays a pivotal role in ensuring that our cities are not only smart but also green and vibrant havens for all residents.

Chapter 9: Environmental Monitoring and Management

As cities around the globe strive to become more sustainable and livable, advanced technologies are playing an increasingly crucial role in environmental monitoring and management. AI is at the forefront of this transformation, providing innovative solutions to combat pollution, optimize waste management, and efficiently manage water resources. With AI-driven applications, cities can now deploy sensors and smart systems that offer real-time data and predictive insights, enabling proactive responses to environmental challenges. These tech-driven initiatives not only improve urban air quality and reduce waste but also ensure the responsible usage of vital resources like water, creating healthier and more eco-friendly urban environments. By harnessing the power of AI, cities are not only able to address current environmental issues but also prepare for future demands, fostering a balance between urban growth and environmental stewardship. As we delve deeper into this chapter, the potential for AI applications in environmental domains unfolds with numerous possibilities, inspiring urban planners, policymakers, and everyday citizens alike to envision a greener, smarter future.

AI Applications in Pollution Control

The surge of Artificial Intelligence (AI) in urban management is revolutionizing how cities address one of the most pressing issues: pollution. As urban populations swell, so does the complexity of managing air and water quality, noise pollution, and waste. AI offers innovative tools to monitor, predict, and control these environmental parameters with remarkable accuracy and efficiency.

At the heart of AI applications in pollution control are sensors and data analytics. Modern cities are being outfitted with a network of sophisticated sensors that continuously collect data on air quality, emissions, and particulate matter. These sensors capture vast amounts of data, which AI systems then process to detect patterns and anomalies.

AI's real power lies in its predictive capabilities. By analyzing historical and real-time data, AI algorithms can forecast pollution levels and identify potential hotspots before they become critical. This proactive approach enables city officials to implement strategies such as temporary traffic restrictions or industrial regulation to mitigate potential impacts.

Consider AI's role in improving air quality. Machine learning models can predict air pollution levels based on various factors, including weather conditions, traffic patterns, and industrial activities. These insights inform decisions about when and where to deploy air quality interventions, like the greening of urban spaces or the installation of air-purifying technologies.

Data fusion, which combines multiple data sources, is another area where AI excels. For instance, integrating satellite imagery with ground sensors provides a more comprehensive view of pollution patterns. AI systems apply complex algorithms to this integrated data, offering a granular analysis that was unattainable with traditional methods alone.

The deployment of AI-powered drones exemplifies a more dynamic approach to pollution control. Drones equipped with AI technology can monitor air and water quality over large areas in real-time. They can navigate predetermined paths or adjust their routes based on detected pollution levels, providing a flexible pollution monitoring solution.

Water pollution is another area where AI is making strides. AI systems can track pollutants in rivers, lakes, and oceans, predicting contamination events and enabling timely interventions. These systems often work by analyzing images and sensor data to identify irregularities in the water's composition.

Noise pollution, though less visible, significantly impacts urban living. AI helps cities tackle this issue by analyzing sound patterns and pinpointing sources of excessive noise. AI-driven solutions can suggest remedial measures, such as rerouting traffic or enforcing noise regulations more effectively.

AI's role in waste management, a key factor in urban pollution, is transformative. By applying AI analytics, cities can optimize waste collection routes, reducing fuel consumption and emissions. AI-powered robots and sorting systems are also enhancing recycling efficiency by accurately segregating waste.

The integration of AI in pollution control systems leads to more sustainable urban environments. However, it requires careful consideration of privacy issues, data security, and the ethical implications of widespread surveillance. Balancing these factors will ensure that AI serves the interests of urban dwellers while respecting individual rights.

Collaborations between city governments, academia, and the private sector can accelerate the development and implementation of

AI solutions for pollution control. These partnerships are vital for sharing knowledge, resources, and technical expertise.

In essence, AI applications in pollution control represent a major leap forward in our ability to manage environmental challenges in urban areas. As we continue to refine these technologies, the prospect of achieving cleaner, healthier cities becomes more tangible. By leveraging AI's potential, we can build urban centers that are not only smarter but also more sustainable.

Smart Waste Management Systems

Smart Waste Management Systems are revolutionizing how cities manage their waste disposal by employing AI technologies that offer a smarter, more efficient approach to dealing with urban waste. By integrating IoT sensors into bins and vehicles, these systems provide real-time data on waste levels, enabling optimized collection routes that reduce fuel consumption and lower carbon emissions. AI algorithms can analyze this data to predict peak waste generation times, ensuring resources are allocated where they're needed most. These systems don't just streamline operations—they empower cities to implement sustainable practices, reducing landfill contribution and promoting recycling efforts through intelligent sorting mechanisms. As cities worldwide grapple with growing urban populations, adopting such innovative solutions not only addresses waste challenges but also demonstrates a commitment to environmental stewardship and smarter urban living.

AI in Water Resource Management plays a vital role within the broader scope of environmental monitoring and management, serving as a cornerstone for the transformation of smart waste management systems. Water, an indispensable resource, presents numerous challenges in terms of quality, distribution, and conservation. In recent years, AI innovations have paved the way for addressing these issues more effectively, allowing urban centers to manage resources with precision and foresight.

Harnessing the power of AI in water resource management entails utilizing sophisticated algorithms and machine learning models to optimize various aspects of water conservation and distribution. These technologies are increasingly integrated into smart waste management systems, ensuring that every drop counts. AI-driven models can predict weather patterns and water demand, enabling cities to adjust their water usage intelligently and sustainably.

Predictive analytics, a significant AI capability, is transforming how cities handle their water resources. By analyzing historical data alongside real-time inputs, urban planners can forecast water shortages and surpluses, ensuring a balanced and efficient supply system. This minimizes waste and overuse, critical objectives in achieving sustainable urban environments. Furthermore, AI can identify leaks and inefficiencies within water networks, substantially reducing the loss of this precious resource.

Moreover, AI's role in monitoring water quality cannot be overstated. Chemical pollutants, pathogens, and other contaminants pose serious threats to public health, making real-time water quality assessment a necessity. AI systems equipped with advanced sensors constantly monitor water characteristics, providing immediate alerts to city officials when abnormalities are detected. Such proactive measures not only safeguard public health but also prevent potentially devastating environmental consequences.

Integrating AI into water resource management is not without its challenges. The initial setup and infrastructure investments can be significant, alongside the need for technical expertise to operate and maintain AI-based systems. However, the benefits, which include long-term cost savings, resource conservation, and improved service delivery, far outweigh these hurdles. Policymakers and urban planners must weigh these benefits against the upfront expenditure to fully embrace AI in water management initiatives.

Collaboration is crucial for maximizing AI's potential in this field. Municipalities, technology companies, and environmental organizations can work together to develop customized AI solutions that meet specific community needs. Shared knowledge and resources foster innovation and allow for the adaptation and scaling of successful models across different urban settings. Additionally, public engagement and awareness campaigns can help communities understand the value of AI in water resource management, ensuring widespread support for these initiatives.

AI's ability to integrate seamlessly with other technological systems enhances its utility in water management. Smart waste management systems, for example, benefit from an AI-driven water resource management approach by reducing water waste and optimizing recycling processes. These systems use data analytics to streamline operations, adapt to changing circumstances, and improve environmental sustainability significantly. By prioritizing efficient water use, AI contributes to a reduction in the carbon footprint associated with conventional waste management models.

It's important to recognize that AI in water resource management isn't a one-size-fits-all solution. Each city is unique, with different geographical, infrastructural, and socio-economic conditions. AI offers the flexibility to tailor solutions to regional needs, whether addressing the water scarcity in arid areas or managing the challenges of excess rainfall and flooding in wetter regions. This adaptability ensures that AI can be a powerful ally in diverse environmental contexts.

Despite the transformative potential of AI, ethical and privacy issues must also be considered. The collection and analysis of vast amounts of data raise concerns about how this information is used and who has access to it. Transparency in data handling practices is essential to build trust between the technology providers and the communities they serve. Establishing clear guidelines and policies around data privacy ensures responsible AI implementation that respects citizens' rights.

Looking ahead, the future of AI in water resource management promises even greater advancements. Continuous innovation will likely lead to more refined algorithms and sensors, increasing accuracy and efficiency. As AI technologies become more sophisticated, they will provide even deeper insights into water usage patterns and environmental impacts, aiding decision-makers in crafting robust policies for water conservation and management.

Overall, **AI in Water Resource Management** is a beacon of hope for sustainable urban living. By confronting water challenges head-on with cutting-edge technology, cities can foster a future where resources are utilized smartly and responsibly. This innovative approach to managing one of our most vital natural resources epitomizes the broader goal of enhancing urban life through technology, inscribing a legacy of environmental stewardship for generations to come.

Chapter 10: Smart Buildings and Homes

In an era where technology increasingly intertwines with daily life, smart buildings and homes are emerging at the forefront of urban innovation. These spaces are designed not just for comfort but also for extraordinary efficiency, employing AI to seamlessly integrate lighting, climate control, and security systems. Imagine a home that adjusts its temperature based on your daily routine or a building that dims lights to save energy when rooms are unoccupied. This is more than convenience—it's a revolution in how we think about living and working spaces. AI-integrated home automation not only enhances our quality of life but also significantly reduces energy consumption, paving the way for sustainable urban growth. Moreover, by adopting AI-driven security features, these environments promise increased safety and peace of mind for residents. As cities continue to grow, the adoption of smart technologies in our buildings and homes is a critical step toward a future that balances innovation with responsible resource management, ultimately leading to more resilient urban centers that can adapt to the changing needs of society.

AI-Integrated Home Automation

In smart buildings and homes, AI-integrated home automation is rapidly evolving into one of the most transformative aspects of modern living. From voice-controlled assistants like Alexa and Google Home to advanced learning thermostats and automated lighting systems, AI is making everyday tasks not only convenient but also significantly more efficient. It isn't just about novelty; this technology has the potential to redefine how we interact with our living spaces.

The first layer of transformation comes through energy management. Automation systems powered by AI can adjust lighting, heating, and cooling based on real-time data such as occupancy and weather conditions. This enables a home to markedly reduce its energy consumption, aligning with global sustainability goals while also lowering bills. Consider a scenario where the AI learns your daily schedule, adjusting the blinds to harvest natural light only when you're present, or cooling the house just before you arrive. These intelligent adaptations contribute to an overall decrease in energy demand.

Security, another crucial component, benefits tremendously from AI. Smart cameras and doorbells with built-in AI capabilities can discern between different people and objects, drastically reducing false alerts. Facial recognition technology allows these systems to know who's visiting and notify homeowners instantly. This sentient layer of security not only contributes to peace of mind but also drastically minimizes potential security risks.

Another remarkable feature of AI-integrated home automation is in personalized living experiences. AI can recommend music or TV shows by analyzing viewing habits or offer cooking suggestions based on what's available in your smart fridge. It's like living with your own personal concierge, who grows increasingly adept the more you interact.

Conversational interfaces and voice automation make these tasks even more seamless. Whether adjusting the lights, setting the thermostat, or even making coffee, speaking a command can initiate actions across various connected devices. This hands-free experience not only appeals to tech enthusiasts but also aids those with mobility issues, making intelligent homes more accessible.

Moreover, AI-driven predictive maintenance features in smart homes can identify potential issues before they manifest into real problems. For example, an AI-integrated HVAC system can alert homeowners to potential malfunctions by detecting irregular patterns in usage and performance. This proactive management reduces costs associated with emergency repairs and extends the lifespan of appliances.

AI also enhances sustainability efforts within the home by managing resources more judiciously. Smart irrigation systems can gauge soil moisture and weather forecasts to water gardens only when necessary, helping conserve water. In urban areas where resources might be strained, such innovations are crucial in maintaining balance between modern living demands and environmental stewardship.

The social implications are just as compelling. With the predicted rise in urban populations, more compact and automated living can help manage the limited space efficiently. Multi-resident buildings, for instance, could utilize AI to balance elevator traffic, optimize parking space usage, or even regulate communal resources like heating.

Yet, the journey towards completely AI-integrated homes isn't without hurdles. As homes become smarter, and more connected, they also become more vulnerable to cyber threats. Security frameworks must evolve in tandem with technological advancements to protect these digital networks from breaches. Personal data privacy, too, is a realm of concern, urging developers to build ethical and secure AI systems.

Despite these challenges, the trajectory toward greater home automation using AI is only accelerating. Companies and developers are investing in robust research and development to overcome existing barriers, with breakthroughs in machine learning, quantum computing, and IoT integration expected to open new possibilities.

AI-integrated home automation holds the promise of revolutionizing personal and communal living spaces. By intelligently managing resources and personalizing experiences, such smart systems are expected to usher in a new era of comfort and efficiency. As cities continue to modernize, aligning smart home technology with broader urban systems will play a key role in the ongoing evolution of urban life.

In essence, AI's influence on home automation extends far beyond individual convenience; it's crafting the blueprint for cities of the future. By driving smarter urban development and fostering sustainable practices, AI isn't just an ancillary tool—it's a cornerstone of transformative urban living that awaits us.

Energy-Efficient Building Designs

Imagine a world where buildings not only meet the needs of their occupants but also actively contribute to environmental sustainability. Energy-efficient building designs are at the forefront of this movement, leveraging cutting-edge technologies to minimize energy consumption while maintaining comfort and functionality. By integrating smart sensors and AI-driven systems, these buildings adapt in real time to changes in occupancy and weather, optimizing heating, cooling, and lighting systems. This not only results in significant cost savings but also reduces the carbon footprint of urban environments. The seamless integration of renewable energy sources, innovative insulation materials, and intelligent design strategies further enhances the sustainability of these structures. As cities continue to expand, embracing energy-efficient building designs is crucial to creating resilient urban landscapes that harmonize technology with nature. This represents a transformative shift in how we conceptualize and construct our living and working spaces, ultimately paving the way for smarter, more sustainable cities worldwide.

Enhancing Security in Smart Homes has emerged as a critical consideration in designing energy-efficient buildings. As smart homes become more integrated into the urban landscape, they present unique challenges and opportunities for enhancing security while maintaining energy efficiency. In the hustle to construct smarter, more sustainable homes, it's easy to overlook the complexities of securing these spaces. Yet, security and energy efficiency don't have to be at odds; instead, they can complement each other to create safer and more efficient living environments.

The advent of AI and the Internet of Things (IoT) has revolutionized how we think about home security. These technologies enable smart homes to learn from the residents' habits and adapt to changes, reinforcing security without compromising energy efficiency. For instance, smart lighting systems can deter burglars by simulating human presence through randomized lighting patterns, reducing the need for constant manual intervention. Integrating these with smart sensors and cameras provides comprehensive security coverage, adapting to real-time situations while ensuring minimal energy usage.

Consider how AI-driven systems can enhance energy management systems within a smart home. Smart thermostats, for instance, can optimize energy consumption based on historical data, weather forecasts, and occupancy patterns. These same systems can enhance security by notifying homeowners of unusual activity, like a window left open during an absence, potentially signaling a security breach. By combining energy management with critical security functions, these technologies create a harmonious balance between energy savings and unmatched vigilance.

Security enhancements in smart homes also extend to digital safety measures. As smart homes rely extensively on connected devices, securing the data that flows through them becomes paramount. Employing AI-driven cybersecurity solutions helps detect and mitigate potential threats before they compromise a home's digital infrastructure. For instance, AI can identify anomalies in data traffic, alerting homeowners or system administrators to unauthorized access attempts, all while learning from each incident to strengthen future defenses.

Another strategic element in enhancing smart home security is the use of biometric technology. These features, like facial recognition or fingerprint sensors, allow personalized access to homes, eliminating the risks associated with traditional keys. Biometric systems can be energy-efficient by powering down unless activated by an approved user, thereby maintaining secure and energy-conscious operations simultaneously.

Moreover, smart home technology can facilitate community safety beyond just individual households. Neighborhoods equipped with interconnected smart security systems can share real-time alerts about suspicious activities or environmental hazards, such as fires or gas leaks. This networked approach not only enhances security at a communal level but optimizes energy consumption across multiple homes by integrating systems for mutual benefit.

However, the integration of security and energy efficiency in smart homes is not without challenges. Issues like data privacy, potential hacking, and the need for ongoing system updates require vigilant attention. Homeowners must be educated and proactive about these risks, ensuring that security systems are regularly updated and that privacy settings are aligned with individual comfort levels.

Implementing robust encryption protocols and ensuring data is stored securely can mitigate many digital security concerns. Regular software updates and partnerships with cybersecurity experts are essential to maintaining the integrity of smart home systems. Additionally, governments and industries must collaborate to establish regulations and standards that protect homeowners while fostering innovation.

In the pursuit of smarter, energy-efficient homes, we must not overlook the importance of designing them with robust security architectures from the ground up. As AI and IoT continue to evolve, so too will the capabilities and complexities of home security systems. By maintaining a focus on both security and energy efficiency, smart homes will not only become paragons of sustainable living but also sanctuaries of safety and well-being.

In sum, enhancing security in smart homes is a multifaceted challenge that requires a careful blend of technology, education, and regulation. By leveraging AI and IoT, we can create environments that are not only energy-efficient but also profoundly secure. As we advance into an era dominated by smart technology, the successful integration of these systems will stand as a testament to what is possible when innovation meets responsibility, driving us toward a future where security and sustainability coalesce in smart and meaningful ways.

Chapter 11: Mobility and Autonomous Vehicles

The transformative power of AI in urban mobility is both vast and profound, reshaping how we think about transportation within cities. As autonomous vehicles progress from concept to reality, they promise to redefine traffic dynamics and urban design. These advancements in automated transit options not only aim to make transportation more efficient but also safer by reducing human error, a leading cause of accidents. Moreover, the integration of smart mobility solutions, including AI-enhanced public transport and ride-sharing services, is positioned to alleviate congestion and reduce pollution, ultimately enhancing urbanscapes for residents and visitors alike. Yet, these technologies also bring challenges, like the necessity for robust infrastructure, policy changes, and ethical considerations in AI deployment. In contemplating the future of autonomous vehicles, cities stand on the cusp of a mobility revolution that promises unprecedented convenience and sustainability, pushing the boundaries of what's possible in urban transportation. Through thoughtful integration and proactive planning, the potential for creating smarter, more interconnected urban environments is immense.

AI in Urban Mobility Solutions

In the bustling world of urban mobility, AI is undeniably a game changer. It's driving the transformation of how people and goods move around cities, making transportation smarter, more efficient, and increasingly automated. Urban environments are being reshaped by AI innovations that promise to address age-old problems like congestion, pollution, and limited accessibility. At the heart of these developments is the potential for AI to revolutionize everything from public transportation management to personal travel experiences.

Picture a city where public transit systems are optimized in real-time based on current traffic conditions, weather patterns, and passenger needs. This is not a distant dream but a reality unfolding thanks to AI. Advanced algorithms process vast amounts of data to predict and respond to congestion, automatically adjusting routes and schedules. These smart systems don't just reduce wait times for passengers; they improve the entire ecosystem by minimizing the environmental impact of urban travel.

AI is also enhancing ride-sharing platforms that operate with unprecedented efficiency. These services rely on AI to connect users with drivers and can predict the best pick-up times and locations, reducing idle time and unnecessary travel. Beyond efficiency, AI-powered ride-sharing fosters inclusion, offering mobility options to those without access to personal vehicles or traditional public transit.

The integration of AI into bike and scooter-sharing programs has also introduced new dynamics into the urban mobility landscape. AI algorithms help optimize the placement and availability of these micro-mobility options, analyzing usage patterns to ensure that people can find rides where and when they need them. This not only broadens access but supports sustainable transport alternatives by encouraging short, eco-friendly trips.

Moreover, AI's role in urban logistics can't be overlooked. Delivery drones and autonomous vehicles, guided by sophisticated AI systems, are beginning to transform how goods are moved within cities. These technologies offer a glimpse into a future where last-mile delivery is not just faster but also has a reduced carbon footprint. As delivery demand increases with the rise of e-commerce, AI is paramount in creating solutions that keep cities moving efficiently.

Mapping and navigation are traditional domains being reinvented through AI. Advanced mapping services provide real-time analysis of various data points, such as traffic flow, road conditions, and even pedestrian movements, to facilitate smoother journeys. AI-enhanced navigation is particularly beneficial in dense urban areas where dynamic rerouting can mean the difference between a quick trip and a frustrating delay.

Sustainability is another crucial aspect of urban mobility where AI is making strides. AI systems enable electric vehicle (EV) fleet management, optimizing charging schedules and routes to ensure maximum efficiency. By predicting energy consumption patterns and electric grid demands, AI helps in designing systems that are both more capable and less taxing on urban resources.

Another promising application of AI lies in autonomous vehicle technology. While fully autonomous vehicles in urban areas present complex challenges, AI is steadily advancing towards making them viable. Self-driving cars can dramatically reduce traffic accidents caused by human error and provide mobility solutions for those unable to drive themselves. As AI continues to evolve, it's a race to balance innovation with safety, ensuring that autonomous systems coexist harmoniously with existing urban infrastructure.

Yet, the introduction of AI in urban mobility also brings challenges. Ethical and privacy concerns are significant, particularly regarding data collection and surveillance. How do cities balance the benefits of AI with the rights of their citizens? Addressing these issues requires transparent policies and inclusive dialogues with the communities that will use and be affected by these technologies.

Furthermore, the digital divide is a critical consideration. As AI technologies become integral to mobility solutions, ensuring all urban residents, regardless of socioeconomic status, have access to these advancements is vital. Bridging this gap might require targeted investments and policy frameworks that prioritize equitable access to smart mobility options.

Economic impacts shouldn't be ignored either. The AI-driven mobility sector is a flourishing field with potential for significant job creation and economic growth. However, it's important to prepare the workforce for potential displacement due to automation. Education and training programs focused on the new skills required for AI-centric environments will be essential for sustaining this transformation.

In conclusion, AI is revolutionizing urban mobility solutions by enhancing efficiency, sustainability, and accessibility. While these developments bring remarkable benefits, they also necessitate careful consideration of ethical, social, and economic factors. As cities navigate this AI-driven future, collaboration among technologists, policymakers, and citizens will be indispensable to ensuring these advances contribute to dynamic, sustainable urban living.

The Future of Self-Driving Cars

Self-driving cars are poised to redefine urban mobility, unlocking a realm where safety, efficiency, and sustainability come together in harmony. As cities evolve, autonomous vehicles present the promise of reducing traffic congestion, lowering emissions, and enhancing accessibility for those previously hindered by traditional transportation methods. With advancements in artificial intelligence, these vehicles are becoming increasingly adept at navigating complex urban environments, learning and adapting in real-time to ensure seamless integration into existing transportation networks. The ripple effects of these innovations are vast, offering transformative opportunities—from decreasing the demand for parking spaces to reshaping urban planning and design. Yet, alongside these exciting prospects, there lie challenges such as ensuring regulatory compliance, addressing ethical considerations, and fostering public trust. As we stand on the brink of this transportation revolution, the potential of self-driving cars to shape smarter, more resilient, and inclusive cities beckons us to reimagine our paths forward, sparking both inspiration and innovation in the urban landscape.

Integrating Autonomous Vehicles into Urban Systems represents a pivotal juncture where innovation meets infrastructure in our rapidly evolving cities. As the allure of self-driving cars captures the imagination of technologists and futurists alike, the practical steps required to seamlessly embed these vehicles into urban ecosystems are monumental. While many envision a future with quiet, efficient, and self-sufficient vehicles navigating our cityscapes, the journey to that reality is filled with both opportunities and challenges that must be addressed methodically.

The foundation of integrating autonomous vehicles (AVs) into urban systems starts with reimagining existing city infrastructures. Most cities were not designed with AVs in mind, and their roads, traffic signals, and signage are optimized for human drivers. Therefore, urban planners face the challenging task of adapting these elements to accommodate not only AVs but traditional vehicles, cyclists, and pedestrians simultaneously. This involves re-evaluating traffic flow patterns, adjusting signal timing, and creating dedicated lanes where feasible. It's not just about placing new technology onto old systems but reshaping those systems to enhance collaboration between human and machine.

Another integral aspect of this integration is communication. AVs rely heavily on real-time data exchange—whether it's vehicle-to-vehicle

(V2V) communication, or interactions with infrastructure (V2I). Imagine a network where traffic lights communicate with vehicles to manage speed and flow or where intersections are equipped with sensors that prioritize emergency vehicles while managing pedestrian crossings. Implementing such technologies requires significant investment and collaboration between public agencies and private tech companies, but the benefits promise increased safety and reduced congestion.

The role of artificial intelligence cannot be overstated in this transformative process. At the heart of AVs, AI systems process vast amounts of data, making split-second decisions that human drivers might not be capable of, or might miss in a moment of distraction. However, this reliance on AI also necessitates robust regulatory frameworks to ensure safety and reliability. Urban systems must evolve to not only accommodate AVs but to harness their AI capabilities to improve public transportation, optimize city logistics, and reduce environmental impacts.

Public transportation systems might initially seem at odds with the rise of AVs, but there's potential for synergy. AVs can complement traditional public transport by facilitating efficient "last-mile" connectivity—bridging the gap between public transit stops and final destinations. This integration could lead to a more seamless, multimodal transportation ecosystem, reducing the reliance on private vehicles while enhancing accessibility across different demographic groups.

The introduction of AVs also opens dialogues about land use in urban areas, specifically concerning parking. If AVs can drop passengers off and then autonomously drive to less congested areas to park or even remain in motion until needed, urban areas could see a significant reduction in the need for parking spaces. This could free up land for other uses such as green spaces, pedestrian walkways, or additional housing, thus contributing to broader urban sustainability goals.

Yet, as we venture into this autonomous era, the ethical and social implications demand our attention. Questions about job displacement within the transportation sector and the digital divide between those who can afford such technology and those who cannot need answers rooted in equity and inclusivity. Moreover, ensuring that AVs are accessible to all urban residents, including the disabled and elderly, is crucial for fostering a cityscape that's as diverse as its inhabitants.

Integrating autonomous vehicles into urban systems is not without its technology-specific challenges, either. Cybersecurity remains a top concern, as AVs must be equipped to fend off potential cyber threats that could compromise the safety of both the vehicles and the city's infrastructure. Rigorous testing and continuous development are non-negotiable as we proceed with deploying AV technology on a large scale.

While these hurdles might seem daunting, the potential benefits call us to action. Reduced traffic incidents, increased transport efficiency, and environmental benefits like lowered emissions make the case for a future where AVs coalesce with urban lifestyles. By championing comprehensive strategies that include a mix of policy innovation, infrastructure investment, and public-private partnerships, cities can become not just smart, but safe and sustainable urban environments for everyone.

Finally, public acceptance and confidence are vital to the success of integrating AVs into urban systems. Effective communication strategies and demonstrable safety records will play important roles in gaining public trust. Public forums, pilot programs, and transparent regulatory approaches will invite citizens to engage with and shape the development of these transformative technologies.

Our cities are poised at the brink of a transformation that could redefine mobility. As we progress down this road, integrating autonomous vehicles into urban systems represents not only a shift in transportation but a concerted leap toward a synthesized urban future. Balancing innovation with inclusivity and safety will enable us to ride forward confidently into this uncharted territory.

Chapter 12: Educational Transformations in Urban Settings

Education in urban areas is experiencing a dramatic shift, largely driven by the introduction of AI technologies. These shifts are transforming traditional educational paradigms, providing opportunities to tailor learning experiences to individual needs. AI-powered platforms offer personalized learning paths, adapting in real-time to students' strengths and weaknesses, and thus enhancing educational outcomes. In these vibrant cityscapes, technology is playing a pivotal role in closing educational divides, especially for underserved communities, by offering accessible and flexible learning solutions. Urban settings, with their diverse populations and unique challenges, serve as fertile ground for developing collaborative learning environments that are both inclusive and innovative. These environments not only foster academic growth but also encourage creativity and critical thinking, preparing students to thrive in the rapidly evolving digital age. As cities continue to evolve, the educational landscapes within them are compelled to innovate, ensuring that knowledge and skills keep pace with technological advancements, ultimately empowering learners and driving societal progress.

AI and Personalized Learning

As we delve deeper into the landscape of urban education, the transformative role of artificial intelligence (AI) becomes particularly compelling. The integration of AI into the educational arena is reshaping learning experiences, especially in urban settings where diversity and resource variability present unique challenges. Personalized learning, powered by AI, offers an opportunity to tailor educational experiences to the individual needs of students, thus maximizing their potential.

Personalized learning might sound like a buzzword, yet it's a cornerstone in redefining modern education. By leveraging AI, educators can now design curricula that adapt in real-time to the cognitive and emotional needs of students. This isn't just about accommodating different learning speeds. It's about fostering an environment where each learner's strengths are harnessed, while their weaknesses are addressed in a manner that's both empathetic and efficient.

The application of AI in personalizing education has opened a new dialogue about the role of teachers. Far from replacing educators, AI serves as an invaluable assistant, offering data-driven insights that can inform teaching strategies. Teachers, thus, can focus more on the human aspects of education—nurturing creativity, critical thinking, and social skills—while AI handles tasks like data analysis and content delivery.

At the heart of AI-driven personalized learning are advanced algorithms capable of processing immense quantities of data. These algorithms can identify patterns in a student's learning behavior, providing insights into areas that require further attention or challenge. For instance, if a student consistently struggles with a particular mathematical concept, the AI system can suggest additional resources and exercises aimed at overcoming this hurdle. Alternatively, it might expedite the curriculum for students who excel, keeping them engaged and motivated.

Moreover, AI facilitates the integration of culturally responsive teaching, which is especially significant in diverse urban environments. By analyzing linguistic, preference, and behavioral data, AI can support teachers in crafting lessons that reflect the cultural backgrounds of their students, thereby fostering an inclusive atmosphere. This customization not only enhances the learning experience but also enriches the school community by celebrating diversity.

Beyond individual classrooms, AI is streamlining educational resource management. Urban schools, often grappling with limited resources, can utilize AI to optimize everything from seating arrangements to instructional materials. By predicting usage patterns and needs, AI ensures that educational tools are allocated efficiently and effectively where they're needed the most, reducing waste and saving costs.

However, as we champion the benefits of AI in education, it's critical to navigate the ethical labyrinth that accompanies such advancements. Data privacy concerns are paramount when dealing with sensitive information about students. There's an urgent need for robust policies to safeguard student data, ensuring that the insights we gain from AI are used responsibly and ethically.

Furthermore, the promise of AI in education shouldn't overshadow the importance of equitable access. As urban schools adopt these technologies, disparities in technological infrastructure and expertise must be addressed to prevent widening the educational gap. Ensuring all students have access to the benefits of AI means investing in technology, teacher training, and community education.

Inspiring examples abound where AI has positively impacted education. Consider a city where dropout rates were significantly reduced after implementing targeted, AI-driven intervention programs focusing on at-risk students. These programs were able to predict dropout risks by analyzing attendance, academic performance, and social factors, allowing schools to intervene early and effectively.

As technology continues to evolve, the future of AI in personalized learning looks promising. There are limitless possibilities for developing more sophisticated tools that further enrich the educational landscape. We envision AI not only in adapting curricula but also in simulating dynamic, interactive learning environments that go beyond traditional classroom settings, offering students hands-on experiences digitally.

As we advance, collaboration among educators, technologists, and policymakers will be crucial. To harness AI's potential fully, urban educational systems must work in synergy to create strategies that address challenges while capitalizing on opportunities for innovation. By doing so, we can build an education system that not only meets the needs of today's learners but also prepares them for the uncharted challenges of tomorrow.

In conclusion, AI's role in personalizing learning within urban settings is multifaceted and transformative. By rethinking how we educate, we're not merely teaching students to adapt to technology; we're empowering them to shape the future of technology through knowledge, creativity, and inclusivity. This journey may be lined with challenges, but it ultimately offers the prospect of an educational system that is as dynamic and diverse as the cities it serves.

Bridging Educational Gaps with Technology

In modern urban settings, technology is playing an increasingly pivotal role in addressing educational disparities. AI stands at the forefront of this transformation, offering tools that personalize the learning experience and cater to diverse student needs. By leveraging data-driven insights, educators can identify areas where students may struggle, creating targeted interventions that enhance learning outcomes. Urban areas, with their unique challenges of diverse populations and resource constraints, benefit immensely from these innovations. Technology fosters a more inclusive educational environment, breaking down traditional barriers and granting students opportunities they might not have had. Furthermore, digital platforms provide access to a wealth of resources, enabling continuous learning outside the classroom. As we harness technology to bridge educational divides, the promise of equitable urban education becomes an achievable goal, inspiring hope for a future where every learner can thrive.

Collaborative Learning Environments can be transformative forces in urban education, where the rapid pace of technological advancement demands innovative approaches to bridging educational gaps. As cities grow more diverse and complex, there is an urgent need for educational systems to adapt, ensuring that all students—regardless of their socioeconomic background—have access to quality learning experiences. Collaborative learning environments, driven by tech innovations, offer promising solutions to this challenge.

At the heart of collaborative learning is the concept of cooperation and shared resources. These environments foster an interactive approach where students and educators actively engage with each other, often alongside technological tools, to enhance learning. In urban settings, where resources might be limited, and classroom sizes can be large, technology-driven collaboration can help create more personalized educational experiences. By using AI-powered platforms, teachers can tailor learning materials to meet the varied needs of their students, breaking down complex topics into accessible chunks.

Utilizing technology in collaborative learning environments not only personalizes education but also provides opportunities for students to connect beyond their immediate urban communities. Online platforms and virtual classrooms foster global interactions, allowing students from different cultural backgrounds to work together on projects and share perspectives. This connection broadens their understanding and equips them with skills necessary for the increasingly interconnected world. In an urban context, where communities are often multicultural, these interactions are critical for fostering inclusion and mutual respect.

Moreover, collaborative learning environments are not restricted to traditional school settings. Urban libraries, community centers, and museums are integrating technology to become hubs of collaborative learning. By providing free access to Wi-Fi-enabled devices and AI-driven educational software, these spaces enable learners of all ages to engage with educational content interactively. Such initiatives also complement school education by offering resources that might not be available in every classroom, thereby addressing some of the gaps inherent in urban education systems.

To illustrate, consider the efforts being made in New York City public libraries, which have transformed sections of their spaces into digital learning environments. This initiative allows students to engage in after-school programs focused on STEM learning, often in collaboration with local universities and tech companies. These programs use AI and other digital tools to enrich the learning experience, offering personalized coaching and interactive lessons that keep students engaged and motivated.

In urban areas, where children might face numerous distractions and challenges, establishing a collaborative learning environment through technology can serve as a gateway to success. It encourages active participation and fosters a love for learning by making education more relevant and engaging. More importantly, by providing equal access to technological resources and collaborative learning opportunities, urban educators can level the playing field, ensuring that all students are prepared for future academic and career challenges.

However, implementing these environments requires careful planning and consideration. Educators and policymakers must address digital equity to ensure all students have the necessary tools to participate fully. Access to high-speed internet and devices remains a barrier for many urban communities. Thus, successful collaboration often requires partnerships between schools, tech companies, and government bodies to provide the infrastructure needed.

Teacher training is another critical aspect of leveraging collaborative learning technologies effectively. Educators must be comfortable using AI-driven tools and integrating them into their teaching strategies. Training programs focusing on these skills are essential, allowing teachers to design and execute lessons that make the most of collaborative opportunities. The shift from traditional teaching methods to technology-assisted instruction might be challenging but is imperative for harnessing the potential of collaborative learning fully.

Research indicates that when students are given the opportunity to work in collaborative environments, their engagement and retention of information improves significantly. Collaborative learning helps develop crucial skills such as critical thinking, problem-solving, and communication. These skills are vital for students who will grow up to navigate the multifaceted challenges of urban living and the globalized job market.

The journey towards creating effective collaborative learning environments in urban education is ongoing. As AI and other technologies continue to evolve, they will undoubtedly offer even more sophisticated tools and platforms that can transform urban education. However, the human element—the relationships between students, educators, and communities—remains at the core of these initiatives. Technology, when used thoughtfully, strengthens these connections, enhancing both the educational experience and the potential for personal growth.

Ultimately, the integration of collaborative learning environments within urban educational strategies is about more than just technology; it's about reimagining what education can be in a modern context. It's about creating spaces where students feel motivated and empowered to learn, where they can be critical thinkers and lifelong learners. This transformation is crucial for cities aiming to become smarter and more inclusive, as education is foundational to every aspect of urban life.

Chapter 13: AI in Cultural and Recreational Spaces

As cities grow smarter, the ripple effects of AI are making waves in cultural and recreational spaces, transforming how we experience creativity and leisure in urban settings. Museums and galleries are embracing AI not just to enhance visitor experiences but to curate personalized art journeys that engage diverse audiences in innovative ways. Meanwhile, smart event management leverages AI to streamline everything from ticketing to crowd control, ensuring seamless and enjoyable public gatherings. AI's role extends into tourism by providing dynamic and tailored experiences, offering tourists a blend of tradition and cutting-edge technology at iconic landmarks and hidden gems. As these spaces evolve, they illustrate the immense potential of AI to enrich urban life, bridging the past with the future while inviting us to continuously reimagine and interact with our environments. These advancements promise vibrant urban landscapes where culture and technology meld seamlessly, crafting experiences that are both deeply personal and universally accessible.

Enhancing Museums and Galleries with AI

In the ever-evolving urban landscape, museums and galleries hold a special place as guardians of culture and art. Integrating AI into these spaces isn't just a futuristic concept—it's a transformation that's already underway, enhancing how we interact with, understand, and preserve art and history. Museums and galleries have long been revered as quiet, contemplative spaces. Now, AI is injecting a dynamic layer of interaction and engagement that broadens their appeal and accessibility.

Imagine walking into a museum and receiving personalized tour recommendations based on your interests. AI technology can now analyze your past engagements, both virtually and physically, to tailor a museum visit like a personal curator, highlighting artworks and installations you'd likely find fascinating. This not only makes every visitor's experience unique but also deepens their engagement and learning, inviting them to explore narratives they might have otherwise overlooked.

Moreover, AI has the potential to break down barriers that exist for many museum visitors. For non-native speakers, AI-powered translation tools can offer real-time translations of exhibit information. Similarly, for individuals with visual impairments, AI can provide descriptive audio guides or even tactile feedback, enabling a multisensory exploration of exhibits. This is a radical shift from the traditional one-size-fits-all approach, making museums more inclusive and diverse in their attendee engagement.

Virtual reality (VR) and augmented reality (AR) are also integral components of this technological revolution in cultural spaces. By using AI to power VR and AR experiences, museums can bring history to life. For example, imagine donning VR goggles to find yourself amidst the bustling streets of ancient Rome or standing beside Leonardo da Vinci as he puts the finishing touches on the Mona Lisa. These immersive experiences can transport visitors to different times and places, providing an experiential context that enriches understanding and appreciation. They're not mere gimmicks but crucial educational tools that make learning history a vivid and engaging experience.

AI's role in curation goes beyond just the visitor's immediate experience; it can also assist behind the scenes in vast, often underutilized archives of museums. AI algorithms can analyze huge sets of data, uncovering trends, themes, or patterns that might go unnoticed by human eyes. This ability helps curators create exhibitions that are not only conceptually robust but also novel and unexpected. Such innovations can lead to new connections and interpretations, continually revitalizing collections and keeping them relevant in today's fast-paced world.

Moreover, AI assists in the crucial task of preservation. With the aid of machine learning, museums can predict deterioration in artworks, enabling timely interventions that could prolong the life of these cultural treasures. Techniques such as object recognition and machine vision can detect subtle changes in artwork conditions that the naked eye might miss, such as fading colors or surface cracks. By identifying potential risks early, AI helps conservators apply preventive measures long before the integrity of the piece is compromised.

Furthermore, as digital collections grow, AI also offers robust archival solutions. With extensive amounts of historical data and high-resolution images becoming available, AI can catalog and classify this massive influx, making it easier to store, search, and retrieve. Proper cataloging not only aids researchers but also benefits educators and students, facilitating access to resources that may not be on display in physical spaces. This digital archiving can also act as a preserving measure against potential physical loss or damage, ensuring perpetuity for future generations.

AI is also reshaping how museums and galleries manage their operational logistics. Predictive analytics powered by AI can optimize visitor flow, reducing bottlenecks and improving the overall visitor experience. By analyzing attendance data, these institutions can forecast peak times and allocate resources accordingly, enhancing visitor satisfaction and operational efficiency. Visitor data can also inform the development of special events or exhibits that cater to audience interests, tying cultural offerings closely to community desires and expectations.

While AI presents myriad possibilities, the human element remains central to museum and gallery experiences. AI serves as a tool that, if wielded with care, enhances rather than diminishes the human aspect of these cultural institutions. Human curators, historians, and educators continue to play vital roles, using AI to add layers of depth and context to their already invaluable work. AI doesn't replace the intricate touch of human insight but augments it, allowing experts to reach and teach audiences in unprecedented ways.

Yet, with AI's profound capabilities comes the responsibility of ethical considerations. The stewardship of cultural patrimony necessitates transparency and caution as AI technologies develop. For instance, how do we ensure AI remains an assistant and not an autocrat in artistic curation? The pressure to balance communities' expectations with AI-generated insights lies at the heart of fostering environments that embrace both innovation and tradition. Cultivating trust between museums, technologists, and the public is paramount to ensuring these spaces evolve into the vibrant, inclusive centers of learning and creativity they promise to be.

The future of museums and galleries in the AI age is an exciting realm of endless potential. As they continue to adopt and integrate AI, these institutions are set to revolutionize cultural education and engagement. They become spaces that not only house art and history but also foster innovation and inspiration. It's a moment of transformation, ready to redefine what it means to experience culture in an urban setting—turning museums and galleries into dynamic spaces where past meets present with the aid of cutting-edge technologies.

Smart Event Management

In the vibrant realm of cultural and recreational spaces, smart event management is revolutionizing the way cities host events. By harnessing the power of AI, organizers can seamlessly coordinate logistics, enhance security measures, and optimize attendee experiences through real-time data analysis. Imagine a concert where AI-driven systems manage crowd flow, adjust sound quality based on environmental factors, and even personalize the music playlist based on audience preferences—all happening without a hitch. AI enables more effective communication and collaboration among vendors, artists, and city officials, making the planning process not only smoother but also more sustainable. The magic lies in AI's ability to predict patterns and adapt dynamically to the ever-changing ambiance of live events, ensuring that each event contributes positively to the urban landscape while delighting its participants. The integration of smart event management illustrates how cities are becoming more intuitive and responsive, setting a new benchmark for cultural engagement and enjoyment.

AI-Powered Tourism Experiences integrate the wonders of artificial intelligence (AI) into the vibrant tapestry of tourism, enhancing cultural and recreational spaces through smart event management. As cities evolve into smart urban landscapes, AI emerges as an invaluable tool for enriching the travel experience, crafting seamless, personalized journeys for tourists while simultaneously optimizing operational efficiencies for event organizers. With the integration of AI, the dynamic realm of tourism is witnessing a transformation that is both exciting and promising.

The ability of AI to tailor experiences to individual preferences is revolutionizing how tourists interact with a city's cultural offerings. Imagine walking through a museum where AI algorithms suggest exhibits based on your previous interests, or strolling through city streets as an AI-driven app offers real-time information about historical landmarks or hidden gems you might otherwise miss. These technologies not only personalize experiences but also make cultural knowledge more accessible, engaging, and relevant to visitors from diverse backgrounds.

In enhancing smart event management, AI can predict trends and optimize resources to ensure smooth operation of events, from small local festivals to international conventions. By leveraging predictive analytics, event organizers can dynamically adjust to variables like crowd size, weather impacts, and resource availability, ensuring that events are not only successful but also sustainable. For instance, AI-driven systems can optimize transportation and accommodation logistics, minimizing environmental impacts while maximizing visitor satisfaction.

Beyond optimization, AI-driven tools facilitate enhanced communication and engagement between tourists and service providers. Chatbots, for example, are widely used in the tourism industry to provide 24/7 support, answering common inquiries, and even handling bookings. This allows for a more efficient guest service experience, reducing wait times and increasing satisfaction. Additionally, these intelligent systems continuously learn from interactions, refining their responses to provide increasingly relevant and personal assistance.

Moreover, AI can enhance security measures during city-wide events, ensuring safety while simultaneously enriching the visitor experience. With AI-powered surveillance and monitoring systems, potential issues can be detected and addressed in real-time, creating a safer environment for local and international tourists. These systems, however, also raise important ethical and privacy considerations that organizers and city planners must address to balance security with individual rights.

The transformative potential of AI in tourism is also seen in the creation of immersive experiences that blend the physical with the digital. Augmented reality (AR) and virtual reality (VR) tools, powered by AI, allow tourists to explore sites in new and interactive ways. For instance, historical recreations via AR can bring ancient sites to life, providing deeper insights and greater educational value. These AI-powered experiences offer tourists novel ways to engage with cultural and recreational content, making tourism more interactive and memorable.

AI-powered systems also facilitate better management of natural resources and tourist flow in popular destinations, mitigating the adverse effects of overtourism. These systems can analyze data to predict visitor patterns, allowing authorities to implement strategies that distribute tourists more evenly across sites and times, reducing strain on infrastructure and preserving local ecosystems. Such proactive measures help maintain the integrity and charm of destinations, so they can be enjoyed by future generations.

While challenges remain, particularly regarding data privacy, inclusivity, and the digital divide, AI offers cities the tools to reimagine tourism in a way that is both innovative and respectful of cultural heritage. Encouraging collaboration between tech companies, government agencies, and cultural institutions is essential for devising frameworks that employ AI to enhance tourism sustainably and equitably. Inclusivity must remain a constant consideration, ensuring that enhanced tourism experiences are accessible to all, regardless of socioeconomic status.

As we continue to explore AI's growing influence on cultural and recreational spaces, it is clear that AI-powered tourism experiences promise not only to enrich the ways in which we explore and enjoy the world but also to ensure that future generations can do the same. By creatively integrating AI into tourism strategies, cities around the world have the opportunity to foster deeper cultural connections, promote economic growth, and enhance the overall quality of life for residents and visitors alike.

Chapter 14: Economic Impacts of AI in Urban Areas

The economic landscape of urban areas is brimming with potential as artificial intelligence (AI) introduces a wave of transformation known to be both promising and complex. As cities embrace AI, the fabric of job markets experiences a reshaping where roles are not only created but also revolutionized, urging the workforce to adapt and innovate. This shift is more than a freeway to job displacement; it's a revival of opportunities that value creativity and critical thinking. Local economies, often struggling under traditional constraints, find new lifelines through AI-enhanced efficiency and novel business models, allowing entrepreneurship to thrive. Communities can witness a rejuvenation of start-up culture as AI lowers the barriers to entry, encouraging innovation and making city landscapes fertile grounds for new ideas. Yet, the pace of these advancements requires balanced strategies, ensuring that the benefits of AI are equitably distributed and that urban infrastructures are prepared for continuous evolution. In this rapidly changing ecosystem, cities stand at the cusp of becoming dynamic hubs where economic resilience is not just an aspiration but a tangible reality, driven by intelligent technologies and forward-thinking policies. The future beckons cities to harness these capabilities effectively, nurturing an environment ripe with economic possibilities while keeping an inclusive mindset at the core of AI integration.

Job Creation and Transformation

As artificial intelligence (AI) weaves itself into the fabric of urban life, it brings with it an evolving job landscape, one teeming with both potential and transformation. It's not just about automation and the displacement of roles that once demanded human intervention; it's also about the creation of new opportunities, necessitating skills that were unheard of decades ago. The integration of AI in various urban sectors is a double-edged sword—while some jobs are at risk of being automated, new and perhaps more enriching roles are being shaped, offering a glimpse into the future of work in cities.

AI technologies are beginning to open doors to novel job prospects. For instance, roles like AI trainers, people who teach machines how to recognize intricate patterns and make decisions, are gaining prominence. Similarly, we're seeing a rise in demand for maintenance experts trained to manage AI systems and ensure their seamless operation. These occupations require a synthesis of technical knowledge and creative problem-solving, bridging a crucial gap as urban environments lean more on AI for efficiency and innovation.

Further, AI's application in urban areas has led to the demand for specialists proficient in ethics and regulatory frameworks, especially as these systems manage sensitive data. Cities that ride the wave of AI integration need professionals who can navigate the moral and legal landscapes wrought by these technologies. As data becomes the new currency, the need for jobs centered around data privacy and security is more palpable than ever. It is these roles that keep the balance, ensuring that the deployment of AI does not trample over the rights and freedoms of city dwellers.

Beyond individual impacts, AI facilitates economic growth, often leading to the emergence of startups and small to medium enterprises that leverage these technologies to innovate and compete. These new businesses create a ripple effect, leading to job creation through trickle-down sectors such as marketing, sales, and customer support. Moreover, urban areas that become hubs for AI-related industries often stimulate employment in construction, logistics, and retail due to increased economic activities.

Yet, the transformation is not without its challenges. One prominent concern is the shift in skill requirements. Many traditional occupations are making way for positions that demand digital literacy and proficiency in complex analytics. There is a pressing need for reskilling programs to help the current workforce transition smoothly into these newly forged roles. Cities must invest in education and training, facilitating collaborations between tech companies and educational institutions to prepare the workforce for this digital metamorphosis.

Public awareness and community engagement are crucial in this transformation. Cities are uniquely equipped to be the frontlines where new job categories are introduced, and citizen awareness campaigns can play a significant role in preparing individuals for this shift. Workshops, seminars, and learning hubs can offer practical insights into AI applications and potential career paths, providing locals the tools to navigate this new world of work.

Additionally, AI is fostering the evolution of jobs that require human intuition and creativity—fields where automated systems cannot replicate human insight. In arts, education, and healthcare, AI provides support, allowing human professionals to focus more on complex decision-making and personal interactions rather than mundane tasks. For instance, teachers are increasingly utilizing AI tools to personalize instruction, tailoring lessons to the needs and learning paces of individual students, thus sparking the need for educators who can seamlessly integrate these tools into their pedagogy.

The transformation of jobs through AI adoption also highlights disparities that cities must address. Not every sector or individual is equally equipped to harness these technologies' opportunities. As a result, city planners and policymakers need to keep inclusivity at the forefront of AI-driven employment strategies, ensuring access to training and development for all demographics, particularly marginalized communities, to prevent exacerbating existing inequalities.

Moreover, AI in urban centers introduces not just new jobs but transformations in existing employment ecosystems. The hospitality and retail sectors, for instance, see shifts toward AI-driven operations, where robotics and automated systems handle standard tasks. Here, the focus shifts toward customer experience roles that necessitate human empathy and interaction, enriching the quality of services provided.

In summary, AI's economic impact on urban job landscapes is multifaceted and profound, presenting challenges and opportunities. While some roles may be at risk, the emergence of new professions centered around AI innovation and maintenance, ethics, and novel technological integrations provides a hopeful outlook. Cities that navigate this transformation effectively, through education, inclusion, and community engagement, will find themselves on the forefront of economic growth and workforce development, paving the way for a more dynamic and resilient urban economy.

Supporting Local Economies with AI

AI's transformative power offers unique opportunities to bolster local economies in urban areas. By optimizing supply chains, it supports local businesses in becoming more competitive and efficient, allowing them to thrive in a rapidly evolving market. Small and medium-sized enterprises benefit from AI-driven insights into consumer demands, enabling them to tailor their offerings and better serve their communities. Moreover, AI tools facilitate more efficient resource allocation, ensuring that local farmers, artisans, and service providers can connect directly with consumers, enhancing market access and fostering economic inclusion. As cities increasingly adopt AI technologies, there's significant potential for job creation within local tech ecosystems, nurturing a new generation of AI specialists who not only drive local growth but also catalyze innovation. Ultimately, the integration of AI helps level the economic playing field, encouraging vibrant, sustainable urban communities where everyone can participate and prosper.

Encouraging Innovation and Startups is vital to ensuring that the economic impacts of AI in urban areas reach their full potential. As cities strive to embrace the myriad possibilities AI offers, fostering an environment where innovation can thrive becomes paramount. The fusion of advanced technology with entrepreneurial spirit drives tangible benefits for local economies, promoting growth and resilience in a rapidly changing world. In this context, AI serves as both an inspiration and a tool for startups, offering innovative opportunities to address urban challenges such as infrastructure, efficiency, and sustainability.

Traditionally, cities have been hubs of innovation due to their density of ideas, talent, and resources. When AI enters the mix, it accelerates the ability of startups to create impactful solutions tailored to specific urban needs. For instance, AI's role in data analysis allows startups to better understand traffic patterns, energy use, and consumer behaviors, leading to products and services that enhance urban living. By leveraging these insights, companies can tailor their offerings to meet the nuanced demands of city dwellers, ensuring relevance and utility.

Moreover, AI-driven startups are at the forefront of creating more sustainable cities. Through innovations in renewable energy systems, smart grids, and waste management, young companies are addressing urgent environmental issues. These startups employ AI to optimize resources, leading to cost savings and reduced environmental footprints. Such efforts contribute significantly to the larger goal of creating smarter, more sustainable urban environments.

The ecosystem surrounding AI innovation is supported by a network of incubators, accelerators, and academic partnerships, which help transform ideas into viable businesses. Urban areas often host these resources, offering startups mentorship and access to capital. Cities benefit by encouraging this kind of entrepreneurial activity, as it not only generates jobs but also attracts investments and talent. Such an ecosystem fosters a vibrant atmosphere where creative ideas can flourish, driving economic growth and diversification.

Access to funding is crucial for the success of AI-driven startups. Venture capitalists and angel investors play a critical role here, identifying promising startups with potential for substantial returns. These investors provide the financial backing required for scaling innovative solutions, ensuring that fledgling companies can move beyond the development phase. When investors see the potential for significant economic impact in urban areas, it encourages a continuous cycle of investment and innovation.

Public policy in urban areas also has a significant impact on the ability of startups to innovate. By crafting policies that incentivize entrepreneurship and offer support structures such as tax breaks and grants, governments can make a tangible difference in fostering innovation. Streamlined regulatory frameworks that protect consumer interests while encouraging technological advancement are essential in this dynamic. When urban policies align with the needs of startups, they create an environment conducive to experimentation and risk-taking, essential components of any innovation process.

AI-driven startups bring a fresh perspective to urban challenges, often approaching problems with agility and creativity. Unlike larger, more established companies, startups are not bound by legacy systems or entrenched processes. This flexibility allows them to pivot quickly in response to new data or changing market conditions, providing them with a competitive edge. In urban contexts, where the pace of change is rapid, such adaptability is invaluable.

Furthermore, these startups frequently collaborate with local governments and organizations in public-private partnerships, leading to solutions that are more aligned with public needs. By working closely with municipal bodies, startups can gain access to proprietary data sets, understanding local issues more deeply. These collaborations often foster relationships that can lead to mutually beneficial outcomes, where public and private interests align towards shared goals, such as improved public services or enhanced citizen engagement.

Educational institutions in urban areas also play a crucial role in fostering innovation by preparing a skilled workforce ready to support AI-driven initiatives. Universities and colleges are increasingly incorporating AI and data science into their curricula, ensuring that the next generation of entrepreneurs is equipped with the knowledge and tools needed to succeed. Partnering with these institutions provides startups with access to not only talent but also research facilities and expert insights, enriching the innovation ecosystem even further.

The social impact of fostering innovation and startups in urban areas extends beyond economic growth. There is a democratization of opportunity as diverse founders from various backgrounds gain access to the resources and support needed to launch and sustain businesses. This inclusivity helps bridge socio-economic gaps and fosters a sense of community among diverse urban populations. It also affords cities the opportunity to tap into a wider array of experiences and perspectives, which is crucial for developing solutions that meet the needs of all residents, not just a select few.

As cities continue to explore the integration of AI into their environments, they must remain committed to supporting innovation and startups. This commitment involves investment in infrastructure that can support high-tech businesses, such as reliable internet and transport systems, as well as policies that encourage experimentation and innovation. By doing so, cities will not only benefit economically but also position themselves as leaders among the smart cities of the future, setting standards for how urban areas can thrive in tandem with technological advancement.

In conclusion, the urban landscape presents a fertile ground for startups aiming to leverage AI for economic and societal benefits. By encouraging innovation, supporting founders, establishing partnerships, and aligning policies with entrepreneurial objectives, cities can effectively harness the transformative power of AI. This approach not only amplifies economic impacts but also enhances the quality of urban life, paving the way for sustainable and vibrant cities of tomorrow.

Chapter 15:
Data Privacy and Security Challenges

As AI continues to transform urban landscapes, the intersection of data privacy and security emerges as a pivotal challenge. Smart cities thrive on data—streams of information flowing from connected devices, systems, and citizens. This data powers innovation, driving efficiencies in transport, healthcare, and governance. But with massive data collection comes the critical responsibility of safeguarding personal information and maintaining trust. Balancing innovation with privacy requires clear strategies that prioritize transparency and consent. Moreover, as cyber threats grow more sophisticated, cities must constantly evolve their defenses to protect against breaches. Addressing these security challenges is not merely a technical necessity; it's essential for fostering an environment where technology enhances urban life without compromising individual rights. Proactive measures and collaborative practices are the keys to ensuring cities remain safe and secure in the face of this digital evolution.

Safeguarding Citizen Data

In the rapidly evolving landscape of smart cities and AI-driven urban environments, safeguarding citizen data has become a vital priority. With cities becoming digital ecosystems that constantly gather, analyze, and utilize data, ensuring this data is protected is crucial. As urban centers become smarter, the amount of data collected from citizens' daily interactions grows exponentially, increasing the potential for misuse if not properly secured. The essence of safeguarding citizen data lies in striking a balance between innovation and privacy—while capitalizing on AI's powerful benefits, cities must equally emphasize robust data protection strategies.

Smart cities leverage a wealth of data from various sources, including sensors, cameras, and mobile applications, to enhance urban living. This data powers intelligent traffic systems, optimizes energy usage, and improves public safety and healthcare services. However, with these advancements come significant privacy concerns. The collection and processing of personal information present risks, such as unauthorized access, breaches, and surveillance issues that could undermine public trust. It becomes imperative for city planners and policymakers to adopt transparent practices and stringent regulations that safeguard citizen data without stifling innovation.

The foundation of data privacy in smart cities begins with the establishment of comprehensive legal frameworks. These regulations should clearly define how data is collected, used, shared, and stored, ensuring that citizens' rights are prioritized over technological capabilities. GDPR in Europe, for instance, inspires global conversations about data privacy, influencing policy in various nations. While legal measures form the bedrock, they should be complemented by a culture of transparency in city administrations. Informing citizens about data handling processes and their purposes nurtures trust and fosters cooperation between citizens and authorities.

Additionally, technology itself plays a crucial role in safeguarding citizen data. Advanced encryption methods and robust cybersecurity defenses form the first line of protection against malicious attacks. Implementing AI-driven security measures can swiftly identify and neutralize threats, minimizing potential breaches. However, these technological solutions must always align with ethical AI standards, ensuring that security measures do not inadvertently infringe on personal privacy. AI can also assist in anonymizing data, stripping personally identifiable information while still enabling useful analysis for city improvements.

Public awareness is another important component in the quest to secure citizen data. Educating citizens about how their data is used, the protections in place, and their rights regarding personal information is essential. By empowering individuals with knowledge, urban areas can encourage citizens to engage actively in discussions and decisions about their data privacy. This approach not only reinforces trust but also integrates community perspectives into data protection strategies, leading to more relevant and accepted policies.

Moreover, partnerships between public agencies and private sector entities can enhance data protection efforts. These collaborations can foster the development of innovative tools and practices that serve both municipal needs and citizen interests. However, public-private partnerships must operate with vigilance to avoid conflicts of interest, and all parties involved should adhere to predefined data governance regulations that maintain citizens' privacy at the forefront.

In this context, cities are beginning to explore decentralized data governance models that empower citizens themselves to control their data. Such models utilize blockchain technology to create a secure, transparent method of managing data permissions. By giving citizens more control over who accesses their data and how it is used, decentralized systems can significantly enhance trust and participation in smart city initiatives.

As urban environments continue to evolve, so too will threats to data security. Urban policymakers must remain vigilant, continuously updating policies and technologies to counteract new and emerging risks. Proactive approaches, such as consistent auditing of data systems and adapting to evolving cybersecurity threats, are necessary to maintain robust safeguards. Additionally, open channels for reporting data-related issues and breaches promote a collaborative defense against potential threats.

The journey of safeguarding citizen data in smart cities will be ongoing. Continuous dialogue among stakeholders, including citizens, technologists, and regulators, is essential in navigating the intricate balance between privacy and innovation. By prioritizing data privacy and demonstrating unwavering commitment to protecting personal information, cities can cultivate an environment where both technological and societal growth coalesce harmoniously. Embracing this ethic will pave the way for smart cities that are not just intelligent but also secure and trusted by all who dwell within them.

Balancing Innovation with Privacy

In the race to embrace cutting-edge technology, cities are navigating the delicate dance of innovation and privacy. On one hand, AI offers tremendous opportunities to transform urban living through smarter infrastructure, public conveniences, and enhanced services. It can predict traffic patterns, optimize energy consumption, and even aid in crime prevention, fundamentally reshaping urban landscapes. But this wave of change brings with it the pressing need to safeguard individual privacy. As data becomes the new currency, cities must balance leveraging personal information for breakthroughs in innovation with respecting citizen rights to privacy. This requires not just advanced technology but thoughtful policy frameworks that build trust and foster transparency. Cities must ensure that innovation doesn't come at the cost of privacy, implementing robust security measures and encouraging public dialogue. By doing so, they can create a future where technology serves the community, ensuring personal data is protected, while urban life flourishes in its new, AI-driven form.

Responding to Cybersecurity Threats is an essential aspect of balancing innovation with privacy in the rapidly evolving landscape of urban AI applications. As cities grow increasingly interconnected, the risks associated with cybersecurity threats become more pronounced. Smart city infrastructures, characterized by complex networks of interconnected devices and expansive data exchanges, often present attractive targets for cybercriminals. The challenge lies in developing robust strategies to protect sensitive data while fostering innovative advancements that can improve urban living.

To effectively respond to cybersecurity threats, municipalities and technology developers must first recognize the diverse nature of potential risks. These threats range from data breaches and ransomware attacks to more sophisticated exploits targeting critical infrastructure like traffic management systems or energy grids. Each type of threat requires a tailored response strategy, underscoring the need for flexibility in cybersecurity planning. By adopting a multifaceted approach, cities can better safeguard themselves against potential vulnerabilities while continuing to harness AI's transformative potential.

Creating a culture of cybersecurity awareness is crucial for any smart city initiative. This involves educating both the public and the workforce involved in maintaining urban systems about potential threats and best practices for mitigating them. Training programs, clear protocols, and access to relevant resources can empower individuals to act as a first line of defense against cyber attacks. Harnessing the power of AI to facilitate real-time monitoring and threat detection is another critical component of a robust cybersecurity framework.

Moreover, regulatory frameworks need to evolve in tandem with technological advancements to ensure they adequately address emerging threats. Policies that mandate regular security audits, data encryption, and robust access controls can help create a more secure urban environment. These regulations must also balance the innovative use of technologies that drive smart city participation with the imperative to protect personal data and privacy. With stringent regulations in place, cities can diminish the risk of data misuse and breach while encouraging technological growth and citizen participation.

Collaborative efforts stand at the forefront of developing effective cybersecurity solutions. Cities must foster partnerships between public entities, tech companies, and academic institutions to share knowledge, resources, and best practices. Collaborative platforms can also facilitate concerted research efforts into emerging threats and build a repository of solutions that cities worldwide can tap into when faced with similar challenges. Such partnerships also offer avenues for small to mid-sized cities to access expertise and resources that might otherwise be beyond their reach.

Additionally, adopting a proactive approach is vital. This includes investing in AI-driven cybersecurity technologies that not only respond to threats but foresee potential vulnerabilities before they can be exploited. AI systems capable of predicting attack patterns and automatically deploying countermeasures can significantly reduce the response time, mitigating damage before it escalates. These technologies can easily integrate into existing urban systems, providing an extra layer of defense that constantly evolves alongside the threat landscape.

Another core consideration is ensuring the resilience of urban infrastructures to cyber attacks. By integrating backup systems, redundant controls, and failsafe mechanisms, cities can maintain operational continuity in the face of a security breach. Developing immersive simulations of potential attack scenarios can help identify weaknesses in existing systems and refine response strategies accordingly. This resilience not only protects physical assets but also instills confidence among citizens that their city is both innovative and secure.

However, no cybersecurity strategy is entirely foolproof. Cyber threats are constantly evolving, and attackers are always finding new ways to circumvent established defenses. Therefore, it's crucial to implement continuous improvement cycles within security measures. By regularly reviewing, updating, and enhancing cybersecurity protocols, cities can adapt to new challenges and enhance their defenses over time. Importantly, feedback from actual security incidents can shed light on overlooked vulnerabilities and contribute to developing more resilient systems.

The future of urban development hinges on the ability to harness AI's transformative potential while ensuring the security and privacy of citizen data. By responding decisively to cybersecurity threats, cities can strike a careful balance between innovation and privacy, building smarter, more sustainable urban environments. In doing so, they fulfill the dual role of being both protectors of public data and pioneers of technological advancement, ensuring that the evolution of urban spaces continues not only unabated but also safeguarded. As we look forward to a future shaped by AI, the commitment to dynamic, robust cybersecurity measures will underpin the success of smart cities worldwide.

Chapter 16: The Role of IoT in Smart Cities

In the ever-evolving landscape of smart cities, the Internet of Things (IoT) stands out as a pivotal catalyst, transforming ordinary urban environments into more connected, efficient, and responsive spaces. Through the seamless integration of sensors and devices, IoT bridges the gap between digital systems and the physical world, enabling real-time data exchanges that streamline city operations. From optimizing traffic flow and enhancing public safety to managing waste and monitoring environmental conditions, IoT applications open up new avenues for sustainable city management. However, alongside these exciting opportunities come challenges related to data privacy, interoperability, and infrastructural demand. By thoughtfully addressing these hurdles, urban planners, policymakers, and technologists can harness the transformative power of IoT to shape cities that are not only smarter but also more livable and equitable for all citizens.

Connecting Urban Devices

In the dynamic environments of smart cities, the Internet of Things (IoT) serves as the connective tissue linking various urban devices, creating an intricate web of communication and functionality. This interconnectedness marks a powerful shift, enabling seamless interactions between previously isolated systems. From streetlights adapting to real-time traffic conditions to waste bins signaling when they're full, the impact of IoT in urban settings cannot be overstated. As cities grow, so does the demand for more agile and efficient infrastructures, and IoT lies at the heart of this transformation.

At the core, urban IoT is about smart choices, not just smart objects. Consider how smart meters in homes can drastically alter energy consumption patterns by providing detailed feedback to residents. It's a change of habit fueled by real-time information. Similarly, in public settings, IoT-enabled devices help cities manage resources efficiently—be it optimizing water usage in parks or reducing energy waste in public buildings. These interconnected systems not only streamline operations but also promote sustainable urban living.

Moreover, the interoperability of urban devices powered by IoT paves the way for higher levels of transparency and participatory governance. Imagine a network of sensors distributed across the city, capturing data on air quality, noise levels, and traffic congestion. Such data sets open up opportunities for collaboration between city authorities and citizens, fostering a culture where data-driven decisions are the norm. This transparency doesn't just improve trust between citizens and institutions but elevates responsiveness and accountability.

In addition to fostering transparency, IoT in urban landscape ensures safety and security. Surveillance cameras, emergency response systems, and environmental sensors collaborate to provide a secure living experience. For instance, IoT sensors can detect unusual levels of pollutants or smoke, triggering timely alerts and responses. In this context, IoT is not just a tool but a guardian of public safety. Whether it's deterring potential threats or assisting in prompt emergency response, interconnected devices play a critical role in the modern urban tapestry.

Nevertheless, as these systems become more entrenched in city infrastructure, the challenges of integrating them effectively come to light. Standardizing protocols for device communication, ensuring compatibility across various platforms, and protecting these networks from cyber threats loom large. It becomes vital to establish robust frameworks that address these challenges head-on. Problems like signal interference, data breaches, and device mismanagement could undermine the potential benefits of IoT, which demands strategic planning and regulatory oversight from city planners.

However, despite these hurdles, the adaptability of IoT systems holds immense promise. By employing machine learning algorithms and AI capabilities, IoT devices can predict and adapt to situations dynamically. Think of autonomous vehicles communicating with traffic lights to optimize travel routes or drones assisting city maintenance by checking hard-to-reach infrastructure. Such innovations exemplify the limitless possibilities of a deftly connected urban environment. This adaptive nature is key to accommodating the evolving needs of cities and their inhabitants.

Furthermore, IoT-driven smart cities are more adept at responding to crises. From flood sensors placed in strategic locations to networks monitoring structural integrity after natural disasters, IoT enables a city to react swiftly and efficiently. Early intervention not only saves resources but also lives. By embracing these high-response systems, cities can dramatically improve their resilience against unforeseen events, positioning themselves as pioneers in disaster resilience.

The financial implications of connecting urban devices also merit discussion. Implementing IoT solutions can lead to significant cost savings through increased efficiency and reduced resource wastage. Cities find they can do more with less, funneling saved resources into improving other aspects of urban life like education, healthcare, and recreational facilities. While initial setups might demand considerable investment, the long-term economic benefits often outweigh these costs.

On a societal level, IoT facilitates inclusivity and accessibility in smart cities. Making urban devices accessible to differently-abled individuals, whether through voice-controlled systems or adaptive interfaces, fosters an inclusive environment. When cities commit to accessibility, they enhance the quality of urban life for all their residents, ensuring that technological advancements serve every sector of society.

In closing, the role of IoT as the backbone of smart city development is pivotal. It reshapes urban landscapes by fostering connectivity, promoting sustainability, and enhancing the daily lives of city dwellers. While challenges persist, they are outshined by the profound benefits and opportunities that arise from such a connected setup. It's a visionary leap into the future—a future where smart cities aren't just a goal, but a reality, continuously evolving with every urban device that comes online.

Benefits and Challenges of IoT Integration

As cities evolve into smart cities, the integration of the Internet of Things (IoT) becomes both a transformative boon and a complex challenge. IoT promises numerous benefits, such as heightened efficiency in urban services, improved quality of life, and sustainable resource management. From smart streetlights that reduce energy consumption to connected waste bins that streamline collection routes, IoT has the potential to significantly enhance urban life. Yet, along with these advantages come notable challenges. The vast amounts of data generated by IoT devices necessitate advanced management and security measures to protect sensitive information and avert potential cyber threats. Compatibility issues across different technologies and the substantial costs of implementation can also impede progress. Moreover, ensuring equitable access to IoT benefits remains a significant hurdle, with the risk of exacerbating the digital divide. Navigating these intricacies is essential for cities to fully harness the power of IoT, paving the way toward future-proof, smart urban ecosystems.

Innovations in Device Communication play a pivotal role in the ever-evolving landscape of smart cities. As industries and governments invest in the Internet of Things (IoT), the way devices communicate with each other fundamentally reshapes urban environments. These innovations not only create efficient systems but also present unique challenges that must be addressed. The rapid advancement of smart technologies leverages interconnected devices to transform city infrastructure, offering unprecedented benefits while posing significant hurdles.

The advent of IoT has revolutionized device communication by enabling seamless exchange of information across various platforms. This connectivity facilitates real-time data sharing, which can drastically improve urban living. With devices communicating efficiently, cities can optimize traffic management, enhance public sanitation, and streamline emergency responses. This network of devices works synergistically, creating intuitive urban ecosystems that respond dynamically to the needs of the city and its inhabitants. IoT innovations, therefore, are not just about connecting devices but are about orchestrating a symphony of technology to improve daily life.

However, the integration of IoT within smart cities is far from straightforward. One of the substantial challenges lies in ensuring interoperability among a diverse range of devices and systems. Manufacturers often produce devices on disparate platforms using different standards, which can lead to communication barriers. Solving this compatibility puzzle is essential for the full potential of IoT to be realized. For effective communication, devices must adhere to standardized protocols, allowing them to exchange data seamlessly and function as intended in a cohesive network.

Security remains another pressing concern in the realm of IoT device communication. As devices increase connectivity and data exchange, the risk of cyber-attacks also escalates. Safeguarding sensitive information involves implementing robust security measures, which require constant updates and vigilant monitoring. Innovations in encryption technologies and secure data transmission are vital to protect urban infrastructure from vulnerabilities. Without strong security protocols, the benefits of interconnected devices could quickly turn into liabilities, jeopardizing privacy and public safety.

Furthermore, innovations in IoT communication have facilitated the development of low-power wide-area networks (LPWANs), specifically designed to connect devices over long distances with minimal power consumption. LPWANs are essential for smart cities because they enable devices to remain powered longer without frequent battery replacements. They also support a vast number of connected devices, making them an ideal solution for large-scale urban applications. The implementation of such networks demonstrates a shift towards sustainable IoT solutions that prioritize energy efficiency and longevity.

The arrival of 5G technology also marks a significant milestone in the enhancement of device communication. 5G networks promise faster data transfer speeds, lower latency, and increased reliability, which are crucial for the successful implementation of IoT systems. With 5G, devices in smart cities can communicate almost instantly, allowing for real-time analysis and decision-making. This technology sets the foundation for more sophisticated applications such as autonomous vehicles and drone deliveries, which rely heavily on swift and dependable communication networks.

Despite these advances, the sheer volume of data generated by IoT devices creates a challenge in terms of processing and management. Edge computing emerges as a solution, bringing data processing closer to where it is generated rather than relying on centralized data centers. By processing data at the edge, latency is reduced, and communication speed is enhanced. This approach not only optimizes the functionality of IoT devices but also alleviates the burden on cloud infrastructure, ensuring smoother operation across the city's network.

Looking beyond technical challenges, there is a critical need for comprehensive policy frameworks to guide the integration of IoT within smart cities. Governments and regulatory bodies must collaborate with technology providers to establish clear guidelines that ensure ethical use and development of IoT technologies. The regulatory landscape needs to evolve alongside technological advancements to address issues such as data privacy, security concerns, and the equitable distribution of IoT benefits among urban populations.

In conclusion, the ongoing innovations in device communication underscore both the possibilities and challenges of IoT integration in smart cities. These advancements pave the way for more intelligent, responsive environments that cater to the needs of their citizens. However, the complexity of these systems necessitates careful consideration of compatibility, security, energy sustainability, and regulatory governance. By addressing these challenges head-on, we can harness the full potential of IoT technology to forge resilient, vibrant cities that enhance quality of life and foster sustainable development for future generations.

Chapter 17: AI Ethics and Regulation

In our journey toward smarter cities, ensuring that AI develops within a robust ethical and regulatory framework is crucial. It's not just about innovation but about creating systems that prioritize human well-being and fairness. Developing ethical AI frameworks involves a delicate balance of promoting technological advancement and safeguarding societal values. Establishing regulatory standards can stimulate responsible AI use, requiring a cooperative effort among government entities, tech developers, and community stakeholders. Globally, perspectives on AI governance differ, but the common thread is the necessity for accountability and transparency. As AI continues to shape urban landscapes, it's imperative to establish guiding principles that prevent misuse and promote equitable benefits for all urban dwellers. The responsibility lies in our collective hands to ensure AI is both a servant to humanity and a guardian of our ethical standards. Insightful regulations coupled with ethical foresight are the twin pillars upon which sustainable and inclusive urban futures can be built.

Developing Ethical AI Frameworks

As artificial intelligence increasingly permeates urban environments, developing frameworks that ensure ethical applications becomes paramount. Cities around the globe are adopting AI technologies to streamline operations, enhance infrastructure, and improve citizen services. With these advancements, however, come questions and concerns about ethical standards. How do cities remain innovative while ensuring the rights and safety of their citizens are protected? Navigating this complex landscape requires a careful, often nuanced approach to balancing innovation with ethical considerations.

At the core of ethical AI framework development lies the principle of transparency. Residents must understand how AI technologies are utilized within their cities and what data is collected about them. For instance, when cities use AI for surveillance through smart cameras or facial recognition, they need to ensure these practices are disclosed to the public. What's more, they must provide information about how the data will be stored, used, and who will have access to it. Transparency not only fosters trust but also encourages public dialogue about acceptable uses of AI, creating space for collaborative decision-making.

Another essential component of ethical AI frameworks is accountability. As AI systems often influence critical decisions, establishing clear accountability structures is vital. Urban planners and policymakers must identify who is responsible if an AI system fails or makes a biased decision. This involves setting up robust oversight mechanisms where actions of AI systems can be audited and reviewed. Furthermore, accountability instills confidence among citizens and can prevent and mitigate the impact of errors or biases that could otherwise erode trust in AI technology.

A major concern within ethical AI frameworks is the potential for inherent bias in AI algorithms. Algorithms are inherently reflective of the data they are trained on, which can inadvertently include societal biases. To mitigate this, it is crucial to ensure diverse datasets that accurately represent the diverse populations they will serve. Researchers and developers must commit to consistently evaluating AI models for fairness and inclusivity, reducing the likelihood of biased outcomes and ensuring equitable treatment across different demographic groups.

Beyond bias, AI systems risk perpetuating existing inequalities if not carefully managed. By prioritizing fairness and equality, ethical AI frameworks can provide checks against such disparities. For example, AI-driven public services should strive to improve access for marginalized groups, ensuring that technology acts as a leveling force rather than an exacerbator of societal divisions. Addressing digital divides is not merely a technical challenge but an ethical obligation cities must approach with sensitivity and a commitment to social justice.

Equally important is the idea of privacy. In the race to smarter cities, the quest for data can sometimes overshadow the individual's right to privacy. Ethical AI frameworks require designing technologies with privacy-protecting features at their core, considering measures such as data minimization, anonymization, and user consent. Citizen privacy should not be sacrificed in favor of technological advancement, but rather technology should evolve to support and enhance privacy rights.

Public involvement in shaping ethical AI frameworks shouldn't be underestimated. Citizen participation provides invaluable insights into community values and priorities. Creating avenues for community engagement and feedback allows frameworks to be reflective of those they serve. Town halls, workshops, and digital platforms are just some ways to invite public discourse, ensuring policies resonate with communities' ethical expectations.

Collaboration between policymakers, technologists, ethicists, and the public is essential in crafting well-rounded ethical frameworks. These collaborations can help bridge the gap between technical capabilities and public acceptability, addressing ethical concerns without stunting innovation. Policymakers should seek expertise from academia and industry to establish guidelines that evolve along with technological advancements. Similarly, the tech industry must be open to regulatory and ethical considerations, exploring ways technology can adapt to meet stringent ethical standards.

As AI technology traverses borders, international collaboration also becomes crucial. Cities, regardless of their locations, face common ethical questions regarding AI use, making global dialogue beneficial. Sharing experiences, best practices, and even failures can guide cities toward better ethical AI solutions, avoiding pitfalls encountered elsewhere. It promotes consistency in ethical standards, enabling urban areas worldwide to adopt technologies with shared values and foresight.

Lastly, continuous evaluation and adaptation present vital elements of ethical AI frameworks. Technology is evolving rapidly, and so should our approach to ethics. Establishing evaluation mechanisms allows for ongoing improvement of AI policies, where learnings from real-world applications can lead to adjustments that better serve citizens. Policies shouldn't remain static, but rather should be living documents that adapt to technological, societal, and ethical changes.

Developing these frameworks is not without its challenges, considering the diverse needs and expectations of urban populations. Nonetheless, the concerted efforts of policymakers, technologists, and citizens can propel cities toward intelligent systems that both respect individual rights and promote communal well-being. By viewing innovation through an ethical lens, urban centers can ensure AI's transformative power is harnessed for all, leading to truly smart and equitable cities.

Establishing Regulatory Standards

In the quest to harness the transformative power of AI for urban development, establishing regulatory standards is essential to ensure these technologies are deployed ethically and responsibly. Crafting these standards involves setting clear guidelines that balance innovation with safeguarding public interest, addressing concerns ranging from data privacy to bias in AI algorithms. By creating robust frameworks, policymakers can provide not only a safer and more reliable AI ecosystem but also foster public trust and facilitate wider adoption. These regulations must be dynamic, adapting swiftly to technological advancements while maintaining a global perspective, considering cross-border ethical implications. It's this delicate balance that will enable cities to leverage AI's potential, driving sustainable growth and improving the quality of urban life for all citizens.

Global Perspectives on AI Governance explore the intricate tapestry of approaches countries and regions adopt to regulate artificial intelligence. In our interconnected world, AI's influence extends across national borders, necessitating collaborative global efforts to establish regulatory standards that balance innovation, ethical considerations, and public safety. These efforts are critical to navigate the dual challenge of harnessing AI's potential while safeguarding against its inherent risks.

Different countries offer varied perspectives on AI governance, heavily influenced by their political, social, and economic contexts. For instance, the European Union has taken a proactive stance in drafting comprehensive regulations with the GDPR serving as a benchmark for data privacy and protection. The EU's AI Act, a pioneering effort, attempts to classify AI systems by risk and sets forth stringent requirements for high-risk systems. This global leadership in AI governance highlights a commitment to developing a trustworthy AI ecosystem that prioritizes fairness, transparency, and accountability.

In contrast, the United States often adopts a more market-driven approach. While there are regulations in place in specific sectors, such as healthcare and finance, the overall regulatory landscape remains somewhat fragmented. Policymakers in the US face the daunting task of balancing innovation with consumer protection, a challenge compounded by the rapid pace of technological advancement. The country relies heavily on industry-led standards and guidelines, which allows flexibility but also raises concerns about self-regulation and the protection of public interest.

China presents yet another distinct perspective. As a global leader in AI development, China's approach combines heavy state involvement with aggressive financing of AI initiatives. The Chinese government has introduced various strategic plans to become the world's leading AI innovator by 2030, focusing on advancing technologies like facial recognition and surveillance systems. However, this approach raises significant ethical questions and concerns regarding individual freedoms and privacy, given the government's emphasis on AI for social control and monitoring.

Regional differences in AI governance also reflect broader cultural and philosophical views on technology and society. Japan and South Korea, for example, demonstrate strong regulatory frameworks emphasizing the human-centric development of AI. Their approaches often integrate ethical considerations with technological progress, embodying a more holistic view of the relationship between humans and machines. This perspective is informed by cultural values that prioritize harmony and cooperation, guiding their respective paths in AI governance.

The global conversation on AI governance is further complicated by the varying levels of technological and economic development among countries. Developing nations face unique challenges as they strive to integrate AI into their urban landscapes. Limited resources, infrastructural constraints, and a lack of skilled workforce can hinder their ability to set and enforce effective regulatory standards. Yet, these nations can leapfrog technological gaps by adopting international best practices and engaging in global dialogue on AI governance.

International organizations play a crucial role in harmonizing AI regulatory standards. The United Nations, through its specialized agencies, works to foster international cooperation and technical standardization. Initiatives like the OECD's AI Principles promote inclusive and sustainable development by outlining ethical guidelines for AI deployment. Such efforts encourage cross-border collaborations and help develop a coherent global framework that supports the equitable and responsible use of AI technologies.

Industry leaders and multi-stakeholder coalitions also contribute to shaping AI governance on the global stage. Tech companies increasingly recognize the importance of participating in regulatory discourse to align their business practices with societal expectations. Collaborative platforms, such as the Partnership on AI, bring together academia, civil society, and industry players to address the ethical implications of AI in diverse applications, from urban planning to healthcare.

The path to effective AI governance is not without hurdles. Geopolitical tensions, differing economic priorities, and cultural disparities pose significant challenges to consensus-building on global regulatory standards. Moreover, the dynamic nature of AI technologies demands adaptive regulatory approaches that can keep pace with innovations while ensuring public trust and safety.

Despite these challenges, a shared commitment to principled AI development offers an optimistic outlook for global governance. The push for AI systems that respect human rights, democracy, and the rule of law necessitates collective action. Open dialogue, mutual learning, and knowledge exchange are essential to crafting regulations that are not only effective but also reflective of diverse global perspectives.

Ultimately, the quest for harmonized AI governance underscores the idea that technology knows no bounds, and neither should our regulatory efforts. As AI continues to redefine urban life, a unified global perspective is imperative to create resilient, smart cities that uphold ethical principles and safeguard the interests of all citizens. Through cooperative governance, we can embrace AI's transformative power while ensuring it serves the broader goal of societal well-being.

Chapter 18: Building Resilience through AI

In an era marked by the increasing frequency of natural disasters and climatic shifts, leveraging AI to bolster urban resilience has never been more crucial. Cities across the globe are embracing AI technologies not just as tools for innovation, but as lifelines capable of predicting, preparing for, and responding to various challenges. AI-driven systems can analyze vast data streams to enhance disaster preparedness, effectively manage resources in real-time during crises, and ensure a speedy, coordinated response. Furthermore, AI aids in developing adaptive infrastructures capable of adjusting to sudden changes, thereby ensuring continuity and efficiency in urban functions. As climate change intensifies, AI is indispensable in crafting sustainable strategies to mitigate its effects, making urban environments both smarter and more robust. Embracing these technologies offers cities a path to greater resilience, safeguarding communities while paving the way for a sustainable urban future.

Disaster Preparedness and Response

In an era where urban areas are becoming increasingly complex and densely populated, disaster preparedness and response are more crucial than ever. The convergence of artificial intelligence (AI) and emergency management offers powerful tools for reinforcing urban resilience against various disasters. Whether dealing with natural phenomena like earthquakes, floods, or human-induced hazards, AI presents novel ways to anticipate, mitigate, and respond to emergencies effectively.

The sheer volume of data generated in modern cities can be overwhelming, yet it holds the potential to revolutionize how urban centers prepare for disasters. AI systems can analyze real-time data from weather stations, satellites, and social media to provide early warning systems with unprecedented accuracy. These systems can predict potential disaster scenarios, offering precious lead time to authorities and communities to strategize and implement emergency procedures. Detection of patterns that foreshadow detrimental events can help prevent loss of life and property.

Take hurricanes as an example. By analyzing vast amounts of meteorological data, AI applications can predict storm paths and intensities with greater precision. This information is vital for orchestrating evacuations and allocating resources efficiently. Instead of relying solely on historical data, AI allows for dynamic modeling that can adjust predictions based on fresh inputs, making forecasts not just more accurate but also more timely.

Moreover, AI doesn't stop at prediction; it plays a crucial role in organizing the response efforts themselves. During disasters, AI-driven tools can optimize logistics to deliver relief supplies swiftly and effectively. Coordinating the movement of people and goods in chaotic conditions is a critical challenge, but AI can manage these dynamics by analyzing traffic patterns and infrastructure status in real-time. Autonomous vehicles and drones, empowered by AI, can be deployed for search and rescue missions or to provide critical supplies to areas that are otherwise inaccessible.

Another innovative application involves the use of AI in monitoring infrastructure resilience. By integrating sensors into key urban structures, AI can assess the stability of bridges, buildings, and other infrastructure elements, alerting authorities to potential weaknesses long before they lead to catastrophic failures. This proactive approach helps cities manage and maintain critical infrastructures, ensuring they remain functional during and after a disaster.

Communication is another critical component improved by AI in disaster scenarios. AI-powered communication platforms can sift through the noise, prioritizing emergency alerts based on severity and reach. In crisis situations where every second counts, such efficient communication can indeed save lives. Furthermore, natural language processing (NLP) technologies facilitate the dissemination of information across multiple languages, ensuring messages reach all demographics within a diverse urban population.

In the context of evacuations, AI supports decision-making by providing insights based on the behaviors and movements of populations. Utilizing AI-driven simulations, authorities can model evacuation routes and strategies, optimizing them for efficiency and safety. AI's ability to process complex variables allows for the development of adaptive evacuation plans that can change based on real-time conditions, ensuring a high level of responsiveness during fluid situations.

However, integrating AI into disaster preparedness and response doesn't come without challenges. Data privacy concerns are at the forefront, requiring careful consideration of what data is necessary to collect and how it is stored and protected. The ethical implications of deploying AI, especially in life-and-death situations, must also be addressed. Establishing trust in AI technologies among the public and ensuring transparent operations will be crucial in maximizing the effectiveness of these systems.

Furthermore, the digital divide poses a significant obstacle. Not all communities will have equal access to AI resources or the connectivity required to benefit from them. Bridging this gap is essential for ensuring equitable disaster response across different socioeconomic backgrounds. Efforts must be made to democratize access to AI technologies, bearing in mind the varied capacities of different urban areas to implement and sustain such initiatives.

Training and workforce development are also essential. Emergency management professionals and city planners must be proficient in AI tools to make informed decisions during crises. Integrating AI education into public service training programs is as vital as the technology itself in building a well-prepared urban environment.

Despite these hurdles, the potential benefits of AI in disaster preparedness and response are substantial and continue to grow as technology advances. Collaborations between AI developers, government agencies, and non-profit organizations are key to designing and implementing effective solutions. Innovation should be driven by a shared vision of creating safer and more resilient cities capable of withstanding the challenges of our unpredictable world.

In conclusion, while AI may not prevent disasters from occurring, it equips cities with the tools to manage them more efficiently and effectively. As urban centers around the world face increasing threats from natural and human-made disasters, the integration of AI into disaster preparedness and response strategies offers a promising path forward. Embracing these technologies with a commitment to ethical considerations and inclusive practices will be integral to building resilient urban futures. Through ongoing innovation and collaboration, AI stands poised to transform the landscape of emergency management, making cities safer for all their inhabitants.

Adaptive Infrastructure

In the ever-shifting terrain of urban environments, adaptive infrastructure stands as a testament to the transformative power of AI. This smart adaptation goes beyond mere structural robustness; it's about a city's capacity to learn, respond, and evolve amidst challenges. AI-driven systems allow infrastructure to anticipate stressors ranging from environmental hazards to urban congestion, dynamically reallocating resources and optimizing performance in real-time. Imagine bridges and roads that autonomously monitor their health, or water systems that adjust flows to prevent shortages. Such intelligent designs not only ensure safety and efficiency but also offer a road map for cities striving to sustain growth against the backdrop of climate change and resource constraints. As cities embrace AI, adaptive infrastructure becomes a beacon of resilience, embodying the fusion of technology and urban development to foster thriving, sustainable urban landscapes.

AI in Climate Change Mitigation As our cities grow and climate change presents new challenges, AI is stepping up as a necessary ally in the development of adaptive infrastructure. The unpredictable nature of our global climate demands flexibility and foresight—qualities that artificial intelligence can provide. When integrated into urban planning, AI can help create infrastructure that not only withstands environmental shifts but also actively contributes to reducing their impact. More than just a tool, AI becomes a strategy for strengthening our cities against climate-related threats while simultaneously reducing our carbon footprint.

Adaptive infrastructure is all about flexibility, pivoting traditional, static designs to more dynamic systems that can adjust to the changing climate. AI's prowess in data analysis and predictive modeling plays a crucial role here. Imagine real-time monitoring systems that can predict weather patterns and adjust public resources accordingly, minimizing damage from potential storms or floods. For example, by integrating AI sensors into stormwater systems, cities can anticipate blockages or overflows and proactively manage resources, thus reducing the risk of urban flooding.

Moreover, AI can optimize the energy consumption of our buildings and transport systems, making them more sustainable and resilient to climate disruptions. Through machine learning algorithms, AI can fine-tune the operation of heating and cooling systems, ensuring energy efficiency without sacrificing comfort. In transportation, AI models can predict peak usage periods and adjust public transportation schedules to reduce emissions while maintaining or improving service levels. These approaches embody the synergy of AI and infrastructure, where both elements inform and enhance each other for more adaptive environments.

Infrastructure isn't solely about huge edifices and sprawling networks; it is equally about understanding and anticipating the needs of those who inhabit our urban spaces. AI excels at processing and interpreting the vast amounts of data generated by city ecosystems. This capability is critical in designing infrastructure that accounts for human activity patterns and environmental conditions. For instance, AI can analyze traffic data to recommend changes in road design or to optimize the timing of traffic signals, reducing congestion and emissions.

One of the most exciting applications of AI in climate adaptation is smart agriculture. As cities expand, they must find clever ways to integrate food production into urban spaces while dealing with the repercussions of climate change. AI-powered systems in urban greenhouses or vertical farms can regulate lighting, temperature, and water usage, dramatically enhancing productivity and resilience against weather variability. By aligning with smart city technologies, AI contributes to urban agriculture solutions that support food security and sustainability.

Artificial intelligence also holds the promise of transforming urban water management. Climate change exacerbates water scarcity and increases the frequency of droughts and floods. Through AI, cities can pioneer smart water systems that monitor consumption patterns, detect leaks early, and efficiently distribute water. These technologies not only conserve resources but also ensure that cities are better prepared for climate-induced water challenges.

Furthermore, AI can play an instrumental role in enhancing biodiversity within urban spaces. By using AI to evaluate habitats and track wildlife movements, urban planners can make informed decisions that protect and enhance biodiversity. Adaptive infrastructure, with AI at its core, supports the development of green spaces that are both wildlife-friendly and beneficial to human well-being, creating urban environments that are more resilient to climate change.

Significantly, AI-driven adaptive infrastructure reflects a collaborative future where technology meets nature. It promotes a sustainable urban evolution by creating infrastructures that mitigate climate change impacts while simultaneously fostering environmentally friendly growth. But to realize this vision, interdisciplinary cooperation is crucial. Engineers, data scientists, urban planners, and policymakers must work together, utilizing AI to craft infrastructure that adapts dynamically to our planet's evolving climate.

AI doesn't just promise to help mitigate the effects of climate change; it challenges us to rethink our approach to infrastructure. We must consider sustainability and adaptability from the ground up, incorporating AI from the planning stage through to development and management. The benefits are clear: infrastructure that not only copes with climate change but thrives and grows stronger in its wake, nurturing resilient, liveable, and sustainable cities that are ready for the challenges of tomorrow.

Chapter 19: Collaborative Innovations and Partnerships

In the ever-evolving realm of urban innovation, the critical role of collaborative efforts between private and public sectors cannot be overstated. These partnerships are driving the smart city movement, creating synergies that harness the strengths of each entity to accelerate technological integration. Public agencies, leveraging their regulatory frameworks and community outreach, blend resources and insights with private sector dynamism and capital, leading to breakthroughs that neither could achieve alone. International cooperation further elevates these projects, sharing knowledge and technologies across borders to address universal urban challenges. By pooling resources, diverse stakeholders enhance their ability to fund and invest in AI technologies, thereby nurturing growth and resilience in urban environments. Such collaborations inspire a shared vision—where cities not only respond with greater intelligence to citizens' needs but also transform sustainably, fostering vibrant, equitable communities for future generations. The journey is not without its hurdles, but through intentional partnerships, the path to smarter cities becomes not just feasible but inevitable.

Private-Public Sector Collaborations

Emerging technologies, especially artificial intelligence, are reshaping how cities operate and evolve. One of the most transformative dynamics in this sphere involves collaboration between private companies and public sector bodies. These alliances are proving critical in the journey toward smarter, more sustainable urban environments, as they bring together the innovation and resources of private entities with the regulatory frameworks and public responsibility of government institutions. Such collaborations are a cornerstone of effective smart city development, helping to bridge the gaps between technological potential and practical urban application.

In many instances, private-public partnerships (PPPs) are driven by a shared vision to enhance urban living conditions. Private firms come equipped with cutting-edge technologies and expertise, such as AI-driven analytics, IoT solutions, and sophisticated data management capabilities. On the other hand, public entities provide infrastructure, policy guidance, and access to a broader community. This combination facilitates the management of critical urban challenges, encompassing everything from traffic congestion and energy distribution to public safety and environmental monitoring.

Several successful collaborations have emerged, demonstrating the tangible benefits of joint efforts. For instance, cities like San Francisco and New York have partnered with tech companies to implement AI systems that optimize traffic flow and reduce congestion. These systems not only improve commute times but also contribute to lowering emissions by minimizing idle times for vehicles. Such initiatives are particularly relevant in densely populated areas where traditional traffic solutions fall short.

Moreover, private-public partnerships are increasingly pivotal in the realm of sustainable energy solutions. Cities are aiming to reduce their carbon footprints and manage energy resources more efficiently.

By collaborating with energy tech firms, municipalities can leverage AI to create smart grids that monitor and optimize energy distribution in real-time. This ensures that energy use is not just sustainable but also economically viable, offering long-term savings for cities and taxpayers while contributing positively to environmental goals.

The benefits of these collaborations extend beyond logistics and infrastructure to the realm of economic growth and social equity. Public sector involvement ensures that advancements are accessible and affordable to all residents, while private enterprises bring in investment and job creation opportunities. These ventures can drive local economies by fostering innovation and providing skills training necessary for emerging job markets.

Of course, it's crucial to address the complexities and potential pitfalls involved in these collaborations. Trust and transparency are fundamental to their success. Public trust can be compromised if citizens feel that data privacy is being sacrificed for technological innovation. Likewise, private companies must be willing to adapt to public sector constraints, like regulatory compliance and the prioritization of public welfare over profit motives.

Successful PPPs are characterized by a mutual understanding of goals and a clear distribution of roles and responsibilities. Stakeholders must engage in continuous dialogue to ensure alignment, adapt to evolving technology and societal needs, and maintain public confidence. Communication also plays a key role in enriching community engagement, fostering a sense of ownership and participation among city residents who are directly affected by these technological changes.

Looking toward the future, private-public collaborations will likely become even more integrated. The growth of smart cities demands that these partnerships not only sustain their momentum but also innovate and explore new models of interaction. Exploring flexible, dynamic frameworks that can adapt to rapid technological changes and complex urban challenges is essential. For instance, developing flexible legal frameworks that protect citizen privacy while enabling seamless data exchange between entities could propel innovation while addressing ethical concerns.

Collaborative platforms that encourage the sharing of lessons learned and best practices can facilitate these partnerships. International cooperation, too, can promote a global culture of knowledge-sharing that transcends borders, opening new avenues for smart city initiatives everywhere. This global perspective not only strengthens individual city projects but also contributes to a collective urban evolution that acknowledges diversity, inclusivity, and sustainability.

Ultimately, the success of smart city initiatives hinges on the strength and adaptability of collaborations between the public and private sectors. These relationships offer a blueprint for cities worldwide to navigate the intricacies of tech-driven urban transformation, balancing innovation with social responsibility. By nurturing these partnerships, cities can not only meet the immediate needs of their communities but also lay down a resilient foundation for future generations, ensuring that urban environments remain vibrant, equitable, and sustainable for years to come.

International Cooperation in Smart City Projects

International cooperation in smart city projects is reshaping urban environments by fostering collaboration across borders. Cities around the globe are no longer isolated in their efforts to become smarter, more efficient, and sustainable. Through sharing knowledge, resources, and technological advancements, municipalities can tackle common challenges such as traffic congestion, pollution, and energy inefficiency. These international partnerships often involve governments, private sector innovators, and research institutions, working together to create frameworks for smarter urban spaces. By learning from each other's successes and setbacks, cities can accelerate progress and implement cutting-edge solutions more effectively. This global collaboration is not just about technological advancement; it's about building resilient and sustainable communities through shared experiences and innovative thinking. The collective effort ensures that the smart city revolution benefits diverse populations and drives the development of urban spaces that are truly prepared for the future.

Funding and Investment in AI Technologies In the realm of smart cities, international cooperation has emerged as a foundational pillar, especially when you consider the intricate web of funding and investment required to drive AI technologies. Across the globe, cities are beginning to collaborate not only in sharing technological insights but also in pooling financial resources to bolster AI innovations that are transforming urban landscapes. This isn't just about economic strategies but about weaving a global tapestry of intelligence and shared progress.

Governments and private entities alike are recognizing that investments in AI technologies extend beyond national borders, particularly for smart city projects that demand sophisticated systems spanning transportation, energy, and security. Various international funding bodies and financial institutions are laying the groundwork for these ambitious undertakings, offering grants and loans that transcend typical geopolitical lines. For instance, the European Investment Bank has been influential in providing financial support to cities integrating AI into urban planning, recognizing the cross-border potential of such innovations.

The flow of funding is contingent on robust partnerships that leverage international cooperation and local expertise. Countries engaging in smart city projects must look beyond traditional funding methods and embrace innovative financial instruments like joint public-private partnerships. These collaborations are essential because they not only provide necessary capital but also align incentives across stakeholders in different countries. Such partnerships are critical in addressing shared urban challenges through AI, be it air quality management or optimizing traffic systems across regions.

Furthermore, the global AI landscape in smart cities isn't limited to developed nations. Emerging economies are increasingly active players, attracted by the potential AI holds for leapfrogging stages of urban development. For international cooperation to be truly effective, funding and investment must consciously include these nations, providing both financial resources and AI-related expertise. This inclusion ensures that smart city technologies promote equitable growth and are not limited to a select few urban regions.

Philanthropic organizations and international consortiums also play a significant role in fostering cross-border partnerships by funding research and development projects aimed at integrating AI in urban settings. These entities often focus on sustainable and socially equitable AI projects, reflecting the broader goals of smart cities. For instance, initiatives led by the United Nations and the World Bank are increasingly incorporating AI for development programs, highlighting the importance of sustainable funding models for technological advancements in urban environments.

Capital allocation for AI in smart cities isn't without its challenges, and careful consideration must be given to ensuring that funding models do not inadvertently widen the technological divide between urban and rural areas, or between different economic classes. Investment strategies must be inclusive, with equity-focused approaches becoming more prominent. Investors are beginning to show interest in ventures that clearly outline their societal and urban benefits, bringing with them mandatory impact assessments as part of the investment criteria.

Moreover, collaboration with international educational institutions can catalyze funding efforts by aligning project developments with academic research and knowledge transfer. Universities and research centers worldwide facilitate collaborative environments where AI solutions for urban challenges can be tested and validated before large-scale implementation. Such collaborations ensure that investments are not just financially sound but also technologically feasible and sustainable over the long term.

Private investments also carry significant weight in the development of AI technologies for smart cities. Venture capitalists are increasingly funneling investments into promising startups that develop AI applications with the potential to scale globally. These investments often dictate the pace of AI innovation, influencing which projects receive priority and, subsequently, which cities emerge as leaders in smart city technologies.

Countries have started to form strategic alliances tailored to harness AI for mutual urban development goals, initiating cross-border projects that blend funding from a variety of sources, including sovereign funds, international banks, and regional development agencies. Such collaborations enable cities to pilot AI initiatives without bearing the full financial risk, encouraging experimentation and innovation across borders.

While the financial infrastructure supporting AI in smart cities is rapidly evolving, transparency and accountability in funding allocation remain critical to sustaining international cooperation. Digital platforms and technological frameworks are being utilized to ensure that financial flows are traceable and that the objectives of funding agencies align with the outcomes of AI projects in urban settings. This transparency bolsters trust, encouraging more partners to engage in cooperative ventures.

Finally, emerging technologies in fintech, including blockchain and smart contracts, are playing a transformative role in facilitating international financial cooperation. These technologies enhance efficiency in managing multi-stakeholder investments, providing more reliable and secure mechanisms for cross-border transactions and partnerships. As funding mechanisms evolve, these innovations are poised to constitute the backbone of future collaborative funding techniques for AI in smart cities.

International cooperation in smart city projects creates a fertile ground for innovation while synergizing funding and investment in AI technologies. By building partnerships across borders, cities can better harness AI to create sustainable, intelligent urban environments that reflect the diverse needs and aspirations of their inhabitants. This complex interplay of funding and investments underscores the value of collaboration in shaping the future of global urban centers.

Chapter 20: Case Studies of Leading Smart Cities

Across the globe, several cities stand out as exemplars in harnessing the power of artificial intelligence to create smarter, more efficient urban environments. From Singapore's seamless integration of sensors and data analytics in urban management to Amsterdam's innovative use of AI for sustainability and citizen engagement, these cities highlight the transformative potential of technology when paired with visionary governance. These trailblazers provide invaluable insights into the successful deployment of AI-driven initiatives, revealing how strategic planning and community involvement can lead to improved living standards, enhanced public safety, and optimized resource management. As cities like Tokyo, Barcelona, and San Francisco continue to push the boundaries of what's possible, they serve as both inspiration and a blueprint for other urban areas aiming to leap forward. Examining their journeys, successes, and challenges offers key lessons for emerging smart cities worldwide, emphasizing the importance of adaptability and collaboration in navigating the complexities of urban innovation.

Pioneering Cities Around the World

In the quest to redefine urban living through technology, certain cities have emerged as pioneers, setting the pace for others to follow. These cities haven't just embraced AI, they've integrated it into the fabric of everyday life. From improving traffic flow to enhancing public safety, these urban centers have taken bold steps to confront modern challenges head-on.

Take, for example, Singapore—an island city-state that's long been at the forefront of technological innovation. Singapore's comprehensive smart city strategy leverages AI across various sectors. Their intelligent traffic management system is a marvel, vastly reducing congestion and optimizing public transportation routes. This commitment to seamless mobility is complemented by strict data privacy regulations that aim to reassure citizens about the protection of their personal information. Here, AI isn't just a tool; it's a partner in urban governance and efficiency.

Turning to Europe, Amsterdam stands out with its use of AI to manage energy resources sustainably. The city's adoption of smart grids has not only enhanced energy efficiency but also reduced carbon emissions significantly. Amsterdam's commitment to sustainability is evident in its decentralized energy management systems, which allow neighborhoods to manage their own resources independently. This local autonomy fosters community involvement and innovation, leading to a more engaged and eco-conscious populace.

In North America, the City of San Francisco has embraced AI to tackle housing challenges. With its high cost of living, the city faces immense pressure to create affordable housing solutions. AI helps urban planners predict growth patterns, allowing them to design infrastructure that supports both density and quality of life. San Francisco's tech ecosystem also ensures a constant influx of AI-driven startups, making it a living laboratory for urban innovation.

Meanwhile, Tokyo illustrates how AI can enhance public safety. By employing AI-driven surveillance systems, Tokyo has significantly improved crime prevention and emergency response. These technologies work in tandem with traditional law enforcement methods to create a comprehensive safety net for residents and visitors alike. Yet, Tokyo has also opened a dialogue about the ethical considerations of AI surveillance, engaging both policymakers and the public in ongoing discussions about privacy and security.

Down under, Melbourne is a stellar example of AI in environmental management. The city's use of AI in water resource management is particularly notable. By predicting usage patterns and weather conditions, Melbourne ensures optimal allocation and conservation of its water resources, a critical concern for a country frequently facing drought. This proactive approach not only preserves resources but also instills a culture of sustainability among its citizens.

On the African continent, Kigali, Rwanda, is a rising star among smart cities. Through strategic partnerships between the public and private sectors, Kigali is harnessing AI to improve urban planning and waste management. Its adoption of drone technology for delivery and data collection is a testament to the city's innovative spirit and its commitment to tech-driven development. As a model for other developing cities, Kigali showcases how emerging technologies can leapfrog infrastructure challenges to elevate living standards.

Venture to the Middle East, and you'll find Dubai, a city that's boldly reimagining its urban landscape with AI. Not just content with having AI-assisted urban management systems, Dubai is looking ahead with its ambitious AI Strategy 2031. This forward-thinking plan aims to make AI integral to city services, from traffic management to legal documentation, positioning Dubai as a central hub of AI research and application in the region.

These pioneering cities share a common vision of using AI to craft more livable, efficient, and sustainable urban environments. They've not only adopted AI technologies but have also built frameworks to support continuous innovation and adaptation. As these cities lead the charge, others look on, drawing lessons from their successes and missteps in equal measure. Their experiences highlight the importance of balancing technological advancement with ethical considerations, regulatory frameworks, and citizen engagement.

Yet, it's not all smooth sailing. Despite technological advancements, challenges remain. The integration of AI into city infrastructures often requires significant investment, and the rapid pace of technological change can outstrip policy updates, leaving gaps that might lead to inequality or exclusion. However, these pioneering cities demonstrate that with strategic vision, adaptability, and collaboration, the path to smart city status is not only possible but also immensely rewarding.

As those leading urban centers continue to explore and innovate, they serve as invaluable case studies for the future of urban living. Through their efforts, they illuminate the transformative potential of AI, setting an inspiring benchmark for cities worldwide to aspire to and learn from. Their stories are not just tales of technological triumphs but also narratives of human aspiration, collaboration, and resilience. In this way, cities like Singapore, Amsterdam, San Francisco, Tokyo, Melbourne, Kigali, and Dubai are shaping the future of urban life, one AI-powered step at a time.

Key Takeaways from Successful Integrations

In exploring the successes of leading smart cities, a few common threads become evident. These cities didn't just leap into the future overnight; instead, they carefully wove technology into the fabric of urban life, ensuring that each step was both deliberate and impactful.

First, the importance of a tailored approach can't be overstated. Successful smart cities didn't just adopt technologies for the sake of being "smart"; they focused on solutions that addressed their unique challenges and goals. Whether it was improving traffic flow, enhancing public safety, or managing energy usage efficiently, these cities took stock of their needs and selected technologies that provided the best fit.

Collaboration emerged as another key factor. Cities leading the smart revolution have often succeeded by forming strong partnerships between public and private sectors. Such collaborations ensure the availability of resources, expertise, and innovation—keeping pace with the rapid advancement of technology. These partnerships also facilitated the sharing of data and insights that helped fine-tune systems to better serve citizens.

Citizen engagement, meanwhile, played a pivotal role. Smart cities thrived when their populations were both informed and involved. By actively engaging citizens in planning, decision-making, and implementation processes, cities saw higher levels of public participation and acceptance. These initiatives helped foster a sense of ownership and pride among residents, ensuring that technological innovations were aligned with community needs and values.

Moreover, adaptability proved essential. As technology evolved, so did the capabilities and strategies of successful smart cities. They remained flexible, willing to adjust plans and integrate new innovations as they became available. This adaptability ensured not only the sustainability of tech initiatives but also kept cities resilient in the face of unforeseen challenges.

Finally, the successful integration of AI and other technologies in smart cities emphasized ethical considerations and data privacy. Ensuring that systems respected individual rights and maintained transparency helped build trust between the city administration and its citizens, a vital component for any smart city endeavor.

These key takeaways illuminate the path forward for cities aspiring to harness technology more effectively. By blending tailored approaches, collaboration, citizen engagement, adaptability, and a commitment to ethics, cities can foster environments where technology enhances urban life seamlessly and sustainably.

Future Prospects for Emerging Smart Cities are unfolding at a pace that few could have anticipated. Fueled by rapid technological progress, smart cities are poised to drastically reshape the way urban areas function. Pioneering cities around the globe are setting benchmarks through successful integrations of AI and other cutting-edge technologies. Yet, as we look toward the future, the emerging smart cities present unique opportunities and challenges. How these cities will integrate technology to build sustainable and people-centered urban environments will greatly influence their success.

The potential for emerging smart cities lies significantly in their ability to embrace an integrated approach to urban living. At the heart of smart city development is the seamless fusion of technology with everyday life. This integration can revolutionize everything—from how waste is managed to how streets are lit—and provide tailored solutions for challenges specific to each city. As technology continues to evolve, it's crucial for emerging cities to focus on collaborations and partnerships that drive this integration forward.

Successful integrations in existing smart cities have shown that public-private partnerships are a critical factor. These collaborations allow cities to innovate more swiftly and effectively than they could independently. In emerging smart cities, building robust alliances with technology providers, government agencies, and community organizations can accelerate their transformation by pooling resources and expertise.

Moreover, the role of data in shaping smart cities cannot be overstated. With sophisticated systems capable of processing vast amounts of information, cities can make informed decisions that enhance efficiency and quality of life. However, this reliance on data brings with it the indispensability of ensuring data privacy and security, a challenge that must be meticulously managed as emerging cities become digitized.

This next wave of smart cities also faces a unique set of economic opportunities. By fostering environments conducive to technological experimentation, emerging cities can become hubs for innovation and startups. Encouraging local entrepreneurship and job creation will be key in ensuring that technological advancements translate into tangible economic gains for residents.

An important factor in the future of emerging smart cities is the growing emphasis on sustainability. As cities expand and populations rise, the pressure on infrastructure and resources intensifies. Smart cities have the potential to address these issues through technology-driven solutions such as AI-powered energy management systems, which optimize usage and reduce environmental impact. By investing in renewable energy sources and smart infrastructure, emerging cities can lead the way in creating more sustainable urban ecosystems.

The human element in smart cities also deserves focus. Technology should be seen as a tool to enhance human interactions and experiences within urban spaces. Emerging smart cities must strive to create inclusive environments that prioritize user-friendly designs and citizen participation. This involves fostering open dialogues with residents to better understand their needs and expectations, thereby ensuring that technological solutions are grounded in real-world applications.

Integration of AI in urban settings presents ethical questions and regulatory challenges. Emerging smart cities will need to navigate the complex landscape of technological governance. Building frameworks that balance innovation with ethical considerations will be paramount. This involves establishing regulatory standards and compliance mechanisms that protect citizens' rights while still fostering technological advancement.

Furthermore, the adaptability of smart cities will be put to the test by unforeseen challenges such as natural disasters. Emerging cities have the opportunity to develop adaptive infrastructures that are resilient to such disruptions. Through the use of AI in disaster preparedness and response systems, cities can become better equipped to handle emergencies effectively, safeguarding both lives and resources.

International cooperation could further fuel the growth of emerging smart cities. By sharing best practices and technologies, cities around the world can learn from one another's successes and failures. This interconnected approach enables cities to develop solutions that are both innovative and culturally sensitive. Such collaborations also offer opportunities to develop and standardize cutting-edge technologies across global urban landscapes, promoting worldwide advancement toward smarter cities.

The evolution of smart cities is not just a technological journey; it is a social and cultural one as well. As emerging smart cities look to the future, understanding and addressing social equity and inclusion will be crucial. Designing cities that cater to diverse populations, while addressing issues of inequality and accessibility, ensures that benefits are shared broadly across all segments of society.

Perhaps one of the most exciting future prospects for emerging smart cities is in the realm of personalization. As AI systems become more adept at processing individual user data, cities will be able to tailor experiences to the needs and preferences of their citizens. From personalized healthcare diagnostics to individually optimized public transport routes, the future landscape of these cities is likely to be intensely personal.

Finally, the future of emerging smart cities is itself an evolving concept. As new technologies emerge and societal needs shift, cities will continue to transform. The key will be maintaining flexibility and a forward-thinking approach. By staying attuned to both technological innovations and human needs, emerging smart cities will be able to pave the way for urban environments that are not only smarter but also more livable and sustainable.

Chapter 21: Community Involvement and Social Equity

As AI continues to redefine urban landscapes, fostering community involvement and ensuring social equity become imperative for sustainable growth. Integrating technology into cities isn't just about advanced systems and automation; it also involves ensuring that every citizen benefits from and participates in the transformation. Civic participation must be encouraged to make smart city initiatives truly reflective of diverse community needs and aspirations. Efforts to mitigate inequality require targeted policies that recognize and bridge the digital divide, ensuring access and opportunities for all. Emphasizing inclusive urban planning can help dismantle systemic barriers and foster environments where diverse voices are heard and valued. By prioritizing collaboration between local governments, communities, and technologists, cities can cultivate a culture of shared vision and responsibility, leading to vibrant, equitable cities for everyone.

Encouraging Civic Participation

In the rapidly evolving landscape of urban environments, civic participation is becoming increasingly vital. It's not just about citizens having a voice; it's about them having an active role in shaping the cities of tomorrow. As smart city initiatives surge forward, they must be grounded in the people's diverse needs and aspirations, whose everyday experiences contribute to the city's fabric. Encouraging civic participation means fostering an inclusive space where technology not only serves but also empowers the community.

Artificial Intelligence (AI) offers unprecedented opportunities to engage citizens in new and innovative ways. From digital platforms that solicit public feedback on urban projects to AI-driven data collection that informs policy decisions, technology can strengthen the channels between citizens and city authorities. But it's not just about technology facilitating communication; it's about shaping the dialogue. Communities need to see themselves as active partners in their city's development if they are to engage fully and inclusively. When people feel their contributions can directly influence outcomes, civic participation doesn't just increase—it thrives.

One way AI can encourage civic participation is through enhancing transparency in governance. Digital tools powered by AI can demystify complex municipal processes and provide clear insights into how decisions are made. Such transparency builds trust, and with trust, citizens are more likely to participate. Digital dashboards, for example, can showcase real-time data on city projects, budgeting, and more, allowing residents to track progress and hold city officials accountable. These dashboards can trigger citizen input on a wide range of issues, from infrastructure changes to public safety concerns.

Moreover, collaborative platforms that leverage AI technology can facilitate community-driven projects. These platforms can gather input from diverse groups and use natural language processing to ensure everyone's voice is heard and considered. This democratization of data and decision-making tools means that neighborhoods can rally around projects that matter most to them, effectively guiding urban planners to prioritize developments that reflect the residents' needs and desires.

Experimentation with participatory budgeting is another promising model for civic engagement. Municipalities worldwide are piloting these initiatives, giving citizens the power to allocate a portion of the city budget to projects of their choosing, enabled by AI systems that can present data in an understandable and actionable manner. This form of engagement not only empowers citizens but also educates them on city governance and budgeting complexities. By directly involving residents in fiscal decisions, cities can become more responsive to the needs of their communities, fostering a greater sense of responsibility and investment among the populace.

Social media, amplified by AI, is also reshaping how civic participation unfolds. AI algorithms can analyze trends and public sentiment to ensure that platforms for civic engagement are both reactive and proactive to community needs. Through social media analytics, city governments can quickly grasp which issues evoke public interest or concern and adjust their engagement strategies accordingly. However, while social media can amplify voices, it also presents challenges like echo chambers and misinformation, which must be mindfully navigated.

Beyond these digital tools, encouraging in-person participation remains crucial. AI can play a role in organizing and promoting community events that foster civic dialogue. Leveraging AI to predict the best times and places for holding town hall meetings or community forums can maximize attendance and engagement, particularly in underserved areas. Bridging the gap between online and offline engagement ensures that civic participation is comprehensive and inclusive, reaching people who may not have digital access.

Education plays a crucial role in encouraging civic participation, too. By integrating AI into educational programs, cities can equip citizens with the knowledge and skills needed to engage meaningfully. Workshops and training can demystify AI and open residents' eyes to its potential impact on their neighborhoods, aiming to build digital literacy from a young age. Such efforts ensure a future generation of informed, active citizens who can wield technology for public good.

However, it's important to address barriers to participation, such as language, accessibility, and technology literacy. AI technologies can bridge some of these gaps by offering translation services and providing platforms that are intuitive and user-friendly. Smart interfaces can be designed to cater to varying levels of tech-savviness, ensuring no one is left behind in the civic dialogue. Engaging diverse communities requires thoughtful inclusivity, recognizing each group's unique challenges and needs.

For civic participation to be genuinely effective, it must be underpinned by a commitment to equity and justice. AI tools should be developed and implemented with careful consideration of biases, especially when they impact decision-making processes. Citizens must have assurances that their participation will not inadvertently perpetuate inequities. Establishing guidelines and frameworks around ethical AI use in public participation can help mitigate these risks, ensuring that technology serves equitable outcomes.

Success in encouraging civic participation will be marked by not only the breadth but also the depth of engagement. When citizens move beyond simply voicing their opinions to actively partaking in decision-making, the relationship between city governments and their residents can be transformed. Authentic engagement fosters a collective civic identity, where diverse communities feel connected to one another through shared goals and aspirations.

The road to increased civic participation is continuous and evolving. As emerging technologies continue to reshape urban landscapes, the opportunities to engage citizens will grow in both scope and creativity. By leveraging AI effectively, cities can cultivate vibrant participatory democracies where citizens are not just passive recipients of urban planning but are co-creators of their community's destiny.

Ultimately, the goal is to build cities that reflect the voices and needs of all their inhabitants. Encouraging civic participation isn't just a strategy—it's a vision for cities where every citizen feels seen, heard, and valued. By integrating AI with intentional community engagement, the future can be one where technology not only enhances urban life but also strengthens the democratic foundations upon which our cities are built.

Addressing Inequality in Smart City Development

As cities embrace smart technologies, it's vital they don't deepen existing inequalities, but rather become platforms for equitable transformation. Developing smart cities with a focus on social equity means ensuring that all communities, including marginalized and underserved groups, have access to the benefits and opportunities these innovations present. This can be achieved by actively involving community voices in the decision-making process, thereby ensuring that technological advancements serve the interests of all residents, not just a privileged few. Prioritizing affordable access to technologies, fostering inclusive design, and investing in digital literacy are essential to blend modern urban advancements with social justice. When cities consciously address inequality, they harness the true potential of smart technologies to strengthen community bonds and improve the quality of life for everyone, paving the way for sustainable, inclusive urban growth.

Building Inclusive Urban Communities rests at the heart of ensuring that smart cities not only thrive technologically but also socially. As we delve deeper into this section, it's critical to remember that addressing inequality in smart city development isn't just about providing everyone access to technology; it's about fostering a sense of belonging in every community. Inclusive urban communities attempt to break down invisible barriers that often restrict access to opportunities, resources, and a voice in city planning.

The move toward smart cities offers a unique opportunity to redefine how urban spaces serve their inhabitants. However, without intentional actions, these advancements risk leaving behind the very people they aim to uplift. Inclusivity must drive the core of smart city frameworks by ensuring that all residents, regardless of socioeconomic status, gender, race, or ability, can access and benefit from technological innovations.

At the ground level, inclusive urban communities focus on participatory governance. This involves integrating public input into planning and decision-making processes. The use of AI-driven platforms can amplify these voices by collecting and analyzing feedback from diverse groups, ultimately shaping urban projects that reflect the community's desires and needs. This sense of engagement not only enhances decision-making but also strengthens the social fabric of cities, creating environments where everyone feels a part of the urban story.

Furthermore, education plays a pivotal role in fostering inclusivity within urban communities. Smart cities must prioritize educational programs that equip residents with the skills needed to navigate and benefit from emerging technologies. By doing so, cities fulfill their promise of equality by breaking the cycle of digital illiteracy and creating pathways to new economic opportunities. Tailored educational initiatives, often powered by AI, ensure that learning is accessible and relevant to all ages, making lifelong learning a central tenet of community development.

Another layer of building inclusive urban communities involves the physical and digital accessibility of infrastructure. City planners and policymakers must prioritize universal design principles that guarantee all spaces are navigable and usable by everyone. This extends to digital spaces, where websites and online services must adhere to accessibility standards. When cities take an inclusive approach, they are not just designing for the minority; they are enriching the experience for all residents.

Moreover, economic disparities can be addressed by leveraging AI and technology to create job opportunities and support entrepreneurial endeavors. Cities can implement programs that nurture local talent, encourage innovation, and provide platforms for small businesses to thrive. This requires a concerted effort from both the public and private sectors to ensure that economic growth is equitable and inclusive.

Cultural sensitivity and inclusivity also play a crucial role in smart city development. Involving diverse cultural perspectives in urban planning fosters environments that celebrate rather than marginalize communities. AI can assist by providing insights into how different demographic groups interact with city spaces, helping to design cities that are vibrant with diverse cultural expressions.

Health equity remains a significant concern within urban environments, and inclusive health systems are paramount. AI-powered health solutions offer great promise for improving healthcare accessibility: mobile health clinics, for instance, can connect underserved populations with necessary services. By integrating AI into public health planning, cities can ensure that healthcare advances are distributed evenly across communities.

Social resilience is another key aspect of developing inclusive urban communities. Resilient cities engage and prepare their residents for future challenges, whether these are related to climate change, economic shifts, or social upheavals. AI can help identify potential vulnerabilities within communities, allowing for proactive measures to strengthen links between residents and local resources. This preparedness fosters a sense of security and cohesion.

Ultimately, the success of building inclusive urban communities lies in continuous collaboration and dialogue among stakeholders. Cities must foster environments where partnerships between governments, private sectors, grassroots organizations, and citizens flourish. These collaborations should aim not only to implement technology but also to ensure that such technology genuinely enhances social well-being and equity.

In conclusion, as we chart the course of smart city development, it is imperative to prioritize inclusivity to avoid increasing disparities. By investing in education, participatory governance, universal accessibility, economic opportunities, cultural sensitivity, healthcare access, and social resilience, cities can build supportive and thriving communities where technology serves as a complement to human connection and equity. Smart cities, if guided by inclusive principles, can truly become the embodiment of innovation that leaves no one behind.

Chapter 22: Challenges and Barriers to Implementation

Implementing AI in urban environments is fraught with a complex web of challenges and barriers that must be navigated to fully realize its transformative potential. Technological hurdles often arise due to the rapid evolution of AI systems and the need for robust infrastructure capable of supporting these advancements. Political and social obstacles further compound these challenges. Gaining consensus among diverse stakeholders, from policymakers to local communities, is essential but often difficult due to differing priorities and levels of understanding about AI initiatives. Additionally, addressing economic constraints is crucial as city budgets are typically stretched thin, making significant investments in AI technologies seem daunting. However, these barriers are not insurmountable. With a concerted effort in building collaborative partnerships, establishing clear policy frameworks, and ensuring equitable resource allocation, cities can move past these challenges and harness AI's potential to create smarter, more sustainable urban spaces.

Overcoming Technological Hurdles

When envisioning a future filled with smart cities, interconnected technologies, and seamless AI integration, it's easy to overlook the substantial technological hurdles that accompany such transformation.

The promise of smarter urban environments relies heavily on our ability to navigate these challenges effectively. Each step towards modernization presents a unique set of obstacles, necessitating a multifaceted approach that combines innovation, strategy, and cooperation.

One of the foremost issues in tackling technological hurdles is the complexity of integrating AI systems with existing urban infrastructure. Cities, particularly older ones, have been built over centuries with layers of developing technology stacked over time. Retrofitting these environments to accommodate new technologies isn't as straightforward as flipping a switch. It requires substantial collaboration among city planners, technologists, and policymakers to ensure that new systems not only fit seamlessly into the existing fabric but also enhance it without causing disruptions.

Equally important is the need for robust and flexible data architectures capable of processing the sheer volume of information generated by AI-powered devices. Cities are data goldmines—with sensors, cameras, and IoT devices collecting vast amounts of data every second. The challenge lies in creating infrastructures that can handle this data efficiently, ensuring swift processing and mining for actionable insights. Without advancements in data storage, transfer, and processing, extracting meaningful benefits from AI technologies could become a bottleneck.

Moreover, the advent of 5G and subsequent technologies promises to alleviate some of these concerns by providing faster and more reliable connectivity. However, the rollout of such infrastructure itself presents a hurdle, particularly in urban areas where logistics, regulatory approvals, and public acceptance complicate deployment. Citywide implementation requires thoughtful planning to avoid interference with existing services and to ensure equitable access across different urban areas.

Security is another formidable challenge in implementing AI technologies in cities. The smarter our urban environments become, the more susceptible they are to cyber threats. Protecting against potential breaches and maintaining data privacy is paramount, especially when it involves sensitive information about citizens. As AI systems increasingly govern critical infrastructure, from transportation grids to healthcare systems, the stakes for cybersecurity grow exponentially. Cementing trust in these systems involves both technical solutions and continual community engagement.

Beyond technological and infrastructural challenges, there exists the hurdle of ensuring skillful human capital capable of driving and managing these advanced systems. There's a growing necessity for education and training programs tailored to equip professionals with the necessary skills to work alongside AI. This skills-gap in the workforce requires bridging through novel educational curricula and upskilling initiatives aimed at fostering the next generation of tech leaders.

Interoperability between technologies also poses a challenge. Different systems and platforms developed by a variety of vendors need to communicate effectively to create seamless outcomes. Establishing common standards and protocols is essential to ensuring that disparate systems can work together harmoniously. Inter-agency collaboration can become vital here, ensuring that as one sector progresses, related sectors aren't left with incompatible technology.

An equally daunting but vital aspect of overcoming technological hurdles is maintaining public trust and ensuring public utility is served. Citizens often express concerns over privacy, loss of control, or technology misuse. Transparent governance and clear communication strategies can help bridge this gap. Engaging the community in decision-making processes and openly discussing the advantages, risks, and ethical considerations upon adopting AI will help assuage fears and bolster participation.

Lastly, the challenge of scalability can't be ignored. A solution that works in one metropolitan area may not be applicable to another, due to differences in scale, culture, or resource availability. Experimentation with pilot projects in smaller areas can offer insights, paving the way for broader implementation. Policymakers should prioritize flexibility in their strategies to allow systems to adapt and evolve with changing urban landscapes and technological advances.

The task of leveraging AI to build smarter, more sustainable cities is vast, but the obstacles, though formidable, aren't insuperable. As novel technologies continue to evolve, so too do the innovative solutions tailored to overcome these technological hurdles. By fostering interdisciplinary collaboration, investing in education and skill development, and remaining transparent and inclusive in technology deployment, the future promises cities that are not only smarter but also more inclusive, responsive, and resilient.

Navigating Political and Social Obstacles

While AI holds transformative potential for urban environments, navigating the political and social landscape presents a distinct set of challenges. Politically, the implementation of AI in cities often encounters regulatory hurdles, as existing governance structures may not be equipped to swiftly adapt to this fast-evolving technology. Policymakers must work diligently to create frameworks that balance innovation with public interest, ensuring that AI solutions are not only effective but also equitable. Socially, the adoption of AI in urban settings stirs concerns around privacy, access, and the potential for technological disparities between communities. Building trust is crucial; cities need to engage citizens transparently in the implementation process, fostering a sense of ownership and shared responsibility. By addressing these obstacles thoughtfully, communities can harness AI's opportunities while minimizing disruption and enhancing social cohesion. The journey is complex, but with strategic collaboration and inclusive dialogue, cities can carve paths toward more intelligent, resilient, and fair urban futures.

Addressing Economic Constraints is a crucial aspect of navigating political and social obstacles in the integration of AI into urban environments. The enthusiasm around smart cities often meets a roadblock of financial feasibility. Just imagining the cost of transitioning a conventional city infrastructure into a technology-driven organism can be daunting. Financial constraints often become the root cause of resistance from governments, communities, and stakeholders, all wary of overextending their financial commitments.

Let's take a closer look at how these economic challenges manifest. For one, the initial capital investment required for AI technologies in transportation systems, energy solutions, and public safety is significant. Cities around the world face the challenging task of allocating funds, either through municipal budgets, subsidies, or private investments, to make these innovations a reality. The high costs are not only associated with acquiring advanced technologies but also with the need for continuous upgrades and maintenance—adding another layer of economic hurdles.

Municipal leaders often grapple with the question: who's going to pay for all this innovation? Affordable short-term financing is rarely available, and long-term debt is unsavory to many who worry about burdening future taxpayers. There's a substantial gap between visionary AI proposals and the financial resources available to implement them. Despite these barriers, cities can't afford to forgo these technologies, as they hold the potential to significantly reduce costs in the long term by enhancing efficiency and resource management.

It's imperative to explore creative financial models to overcome these constraints. Public-private partnerships have emerged as one strategy, inviting investment from corporations eager to showcase their technology and establish footholds in smart urban markets. In some cases, leasing models for technology provide a more palatable alternative to upfront purchases, allowing cities to spread out payments over time while enjoying immediate benefits. This way, economic hurdles can be methodically dismantled without overwhelming municipal budgets.

Incorporating AI in urban settings is not merely about overcoming high costs or finding funding sources. Cities also need to consider the economic sustainability of these initiatives. If AI integrations do not prove cost-effective in the long run, they risk obsolescence or abandonment. Therefore, a keen focus on return on investment and long-term economic viability must underscore every project. By ensuring that projects are economically sound, they gain better traction and acceptance among communities who might otherwise be skeptical of their value.

However, economic constraints aren't only a matter of fiscal planning. Political decision-making is often influenced by competing agendas – from vested interests to strategic city branding efforts. Leaders might face pressure to select initiatives with the most immediate impact, sidelining those that may offer long-term benefits. Here, AI advocates must articulate the long-term value proposition of these investments, which can ultimately lead to greater economic resilience, social benefits, and enhanced citizen well-being.

Furthermore, aligning AI projects with broader economic development goals can help cities secure the necessary funding and public support. For instance, if AI can be harnessed to create jobs, enhance local businesses, or attract new industries, there's a compelling case for its integration. This not only addresses economic constraints but also strengthens the city's growth trajectory, leading to more profound political and social buy-in.

A city's transition to a smarter entity with AI technologies is undoubtedly complex, particularly with contingent economic constraints. But history has shown that economic hurdles can often foster innovation rather than stifle it. Crowdfunding initiatives, urban innovation competitions, and tech incentives exemplify how cities can creatively muster financial resources, align with future-forward policies, and dismantle barriers to AI adoption.

It's critical to assess case studies of cities that have navigated these financial challenges successfully. These examples can serve as blueprints for cities embarking on their own smart city journey, providing insights into innovative financing methods and cost-effective deployment strategies. Additionally, sharing failures can be just as instructive, offering an understanding of pitfalls to avoid.

In summary, addressing economic constraints is a dance between strategic planning, innovative financing models, and long-term economic justification. By approaching these challenges with creative solutions and an eye for sustainable integration, cities can gradually transform into smart hubs of tomorrow. The cost of inaction—remaining tethered to outdated and inefficient systems—underscores the necessity of finding ways to surmount these financial and political obstacles. As cities strive to become smarter, the resolution of economic constraints will determine how effectively they can harness AI to reshape urban life.

Chapter 23: Future Trends in AI and Urban Development

As we look ahead, emerging trends in AI suggest a transformative phase for urban development. The relentless pace of innovation is expected to yield sophisticated AI systems that redefine city living through unprecedented connectivity and efficiency. Autonomous technology will likely play a pivotal role in this evolution, from self-regulating traffic grids to intelligent waste management solutions, paving the way for cities that function like well-oiled machines. Moreover, as AI's capabilities expand, urban planners and policymakers must remain vigilant about ethical considerations and ensuring equitable access to these advancements, so all citizens can benefit. The future city will be one where technology and humanity coexist harmoniously, crafting environments that are not only smart but also sustainable and inclusive. This vision requires continuous adaptation and collaborative efforts to anticipate and address the challenges that lie ahead, aligning technological progress with socio-economic goals. As we prepare for these upcoming transformations, it's clear that AI will be at the heart of reshaping our urban landscapes for the better.

Anticipating Technological Advances

The field of artificial intelligence continues to evolve at a breakneck pace, and its impact on urban development seems limitless. New technological breakthroughs are not just on the horizon; they're already reshaping how we envision cities. As we look ahead, there are several key areas where we can expect AI to revolutionize urban life.

One of the most significant advances will be in the realm of autonomous mobility. Self-driving vehicles are expected to undergo rapid improvements, making them safer and more efficient. These advancements will likely transform public and private transportation networks, reducing congestion and pollution while increasing accessibility for those unable to drive. The ripple effect will touch every aspect of urban planning, from road design to public transit integration. Imagine cities where traffic accidents are rare, and parking lots are a thing of the past. It's not just a dream—it's an anticipated reality as AI continues to advance.

Alongside transportation, AI-driven energy solutions are expected to make cities more sustainable. By optimizing energy consumption and enhancing the efficiency of renewable sources, AI can help urban areas significantly reduce their carbon footprints. Breakthroughs in energy storage technology will allow for better utilization of solar and wind power, even when the sun isn't shining or the wind isn't blowing. Smart grids will become smarter, learning from usage patterns to predict and meet energy demands seamlessly. This transition is not merely an ecological necessity but an economic opportunity, spurring green jobs and technologies.

Communication networks, too, will be profoundly affected. The advent of 6G and beyond will offer unprecedented connectivity, facilitating the real-time data exchange necessary for advanced AI applications. Cities of the future will leverage this connectivity to implement AI-driven solutions that span across sectors, from healthcare to security to commerce. Enhanced communication will support the integration of Internet of Things (IoT) devices, making smart cities more responsive and intuitive. Predictive analytics will become the backbone of IoT systems, anticipating issues before they become problems and allowing for preventive action.

The integration of AI will not be limited to new technologies; it will blend with existing systems to create innovative solutions. For instance, predictive analytics combined with real-time data from interconnected devices can lead to smarter urban infrastructure. Bridges and roads equipped with AI-driven sensors can monitor wear and tear, triggering maintenance before structural issues arise. This proactive approach to urban management will extend the lifespan of city infrastructure and improve public safety.

Moreover, urban areas will see AI transforming social services. AI algorithms are increasingly capable of processing vast amounts of data to identify societal trends and patterns. This capability allows for targeted interventions in areas like public health, education, and housing. Policymakers can use these insights to craft solutions that are both effective and equitable, addressing issues such as inequality and access to services. In essence, AI will become a tool for empowerment, enabling cities to meet the needs of their most vulnerable citizens more effectively.

The advances in AI won't come without their challenges. Ethical considerations will be paramount, especially as AI becomes more entrenched in public life. Cities will need to establish guidelines and regulations to ensure that AI technologies respect privacy and are not used to perpetuate discrimination. There is also the challenge of ensuring public trust in AI systems. Transparency in how AI decisions are made and accountability when things go wrong will be critical to garnering public support. By addressing these challenges proactively, cities can ensure that technological advances benefit everyone.

Another critical consideration is the digital divide. Cities must prioritize equitable access to AI technologies to prevent widening the gap between those with and without digital access. Infrastructure investments must ensure all residents benefit from technological advancements. This will mean prioritizing broadband access in underserved communities and providing education and resources to enable everyone to participate in the digital economy. Only by addressing these disparities can cities fully realize the potential of AI.

Looking ahead, AI's role in crisis management will also become more pronounced. As urban centers face increasing threats from natural disasters and climate change, AI can provide timely and actionable insights. From predicting extreme weather patterns to optimizing emergency response efforts, AI offers tools that can save lives and mitigate damage. Cities equipped with AI-driven systems will be more resilient to future challenges, ensuring urban safety and continuity.

Moreover, AI's capacity for rapid innovation suggests that future developments may address unforeseen challenges. The adaptability of AI systems allows them to evolve alongside urban environments, suggesting a symbiotic relationship where both technology and cityscape influence each other's development. This dynamic will lead to urban areas that are not just smarter but also more adaptable and sustainable in the face of change.

In conclusion, the anticipated technological advances in AI will profoundly transform urban development. From making transportation autonomous and energy consumption more efficient to enhancing communication networks and social services, AI's reach is far. Yet, as these technologies continue to evolve, they must be implemented thoughtfully. There's a need for ethical frameworks, equitable access, and robust public policy to ensure that everyone benefits from AI's potential. As cities prepare for these changes, collaboration between stakeholders—governments, tech companies, and communities—will be crucial. The cities of tomorrow will be testaments to what can be achieved when technology advances thoughtfully and inclusively.

Preparing for Tomorrow's Urban Landscape

The cities of tomorrow aren't just a distant possibility; they're being built today through the innovative use of AI. As urban centers continue to swell with burgeoning populations, the need for intelligent systems becomes apparent in shaping environments that are both sustainable and livable. Forward-thinking urban planners are leveraging AI to anticipate future challenges—be it in resource management or infrastructure resilience—and to craft solutions that embrace growth without compromising quality of life. From reimagining public spaces to integrating autonomous transit, the aim is to create urban landscapes that are not only smarter but also more inclusive and equitable. Embracing AI in urban development means looking beyond mere technological advancements; it's about constructing cities that harmonize with the human experience, fostering communities that thrive amidst change and innovation.

Envisioning the Next Phase of Smart Cities is an exploration into what the future holds for our urban environments as they become increasingly intelligent and interconnected. As we navigate through an era where technology is seamlessly woven into the very fabric of urban life, the possibilities for transformation are endless. This section delves into the emerging trends in smart cities, driven by rapid advancements in AI, that promise to redefine how we live, work, and interact in our urban spaces.

Imagine a city where traffic lights don't just change based on timers or pedestrians at the crossings, but instead, they intelligently adapt to real-time traffic conditions, minimizing congestion and reducing commute times. This vision is becoming a reality as AI powered traffic management systems analyze data from an array of sensors and cameras stark across the urban landscape. These systems not only improve traffic flow but also contribute to reduced emissions and enhanced quality of life for urban dwellers.

Beyond traffic management, smart cities are poised to revolutionize urban living through the development of sustainable and energy-efficient infrastructures. AI-driven predictive analytics are enabling city planners to forecast energy consumption patterns, optimize resource allocation, and integrate renewable energy sources more effectively. This foresight ensures that urban growth is aligned with sustainable practices, facilitating the creation of cities that not only cater to our current needs but also preserve resources for future generations.

Public safety and security are also being transformed through AI. By utilizing machine learning algorithms and vast databases of urban crime statistics, predictive policing is enabling law enforcement to anticipate and strategically respond to criminal activity. These advancements raise important discussions around ethical considerations, privacy, and civil liberties, emphasizing the need for transparent governance and robust regulatory frameworks to balance innovation with the protection of individual rights.

In the realm of healthcare, smart cities are driving a shift toward more personalized and accessible health services. AI is enabling remote monitoring and telemedicine, where patients can receive real-time diagnostics and treatment recommendations without the need to visit a hospital. This not only alleviates the burden on healthcare systems but also ensures that citizens have timely access to necessary care, contributing to healthier, more resilient communities.

As we progress into this next phase, the digital divide remains one of the most pressing challenges. Ensuring that all citizens have access to the benefits of smart city technologies is crucial. AI offers solutions that can bridge these gaps by providing platforms for enhanced public services, including education, transportation, and civic engagement, tailored to meet the diverse needs of urban populations. Technological equity becomes a cornerstone of building inclusive urban communities.

Looking forward, AI's role in urban planning is only set to grow. Advanced simulation models utilizing AI allow planners to visualize and predict urban growth, assessing the impact of infrastructure changes before they're implemented. These tools foster strategic decision-making that prioritizes efficiency, sustainability, and citizen well-being.

The integration of AI in cultural and recreational spaces is another area witnessing remarkable innovation. Museums, galleries, and event spaces are leveraging AI to create immersive and interactive experiences that captivate and educate visitors like never before. These smart transformations not only enhance cultural appreciation but also provide economic opportunities through tourism and local business growth.

Understanding the economic impacts of AI in urban settings is vital. While AI can lead to job displacement in some areas, it simultaneously creates opportunities for new industries and professions, propelling local economies forward. Fostering a culture of innovation and entrepreneurship becomes intrinsic to how cities capitalize on AI's potential.

As we envisage these advancements, data privacy and security challenges loom large. Securing citizen data requires innovative approaches that maintain trust while promoting transparency. It is crucial for smart cities to adopt strategies that balance technological progress with rigorous safeguarding protocols to protect citizens' personal and sensitive information.

Finally, the next phase of smart cities will be shaped by collaborative efforts across borders, sectors, and communities. International cooperation and private-public partnerships are essential in sharing knowledge, resources, and technology. As cities grow smarter, they must also grow more connected, learning and adapting collectively to overcome challenges and harness opportunities.

In conclusion, the vision for the next phase of smart cities is one of optimism and resolute advancement. With the strategic integration of AI, there lies the potential not only for enhanced efficiency and convenience but also for the creation of truly sustainable, equitable, and vibrant urban ecosystems. Enabling this transformation will require a concerted effort from all stakeholders—government, industry, and citizens alike—to ensure cities are prepared to meet the challenges and opportunities of tomorrow's urban landscape.

Chapter 24: Opportunities for Further Research

As we conclude the exploration of how AI is continuously reshaping urban environments, it's clear there's a wealth of opportunities for further research. Identifying knowledge gaps in areas like ethical AI implementation and adaptive urban infrastructures is crucial. Researchers should delve into these realms, ensuring innovations align with societal values and sustainable development goals. Encouraging continued inquiry around the integration of AI into public governance and infrastructure can help tackle emerging challenges of data privacy and social equity. Collaborative research initiatives involving academia, industry, and governments are vital in cultivating diverse perspectives and driving holistic solutions. By fostering a multidisciplinary approach and engaging a wide array of stakeholders, we can ensure that future smart cities not only boast technological advancements but also promote inclusive communities that enhance the quality of urban life for all citizens.

Identifying Knowledge Gaps

As urban areas globally become more entwined with artificial intelligence and related technologies, the identification of knowledge gaps becomes imperative. This is a rich landscape filled with both possibilities and uncertainties, which makes it essential to understand where our knowledge falls short. One immediate area where knowledge gaps are evident is in the long-term societal impacts of AI integration in cities. While we're quite good at forecasting short-term technological advancements, the cascading effects on social structures, civic engagement, and community well-being remain somewhat obscure. Without comprehensive studies that explore these dimensions, we risk developing solutions that are innovative but not wholly beneficial or equitable.

Moreover, the intersection of AI with human behavior in urban spaces is a critical yet underexplored domain. For instance, how do smart city technologies change the way individuals interact with their environment and with each other? The notion of a "smart city" can sometimes be limited to technology infrastructure and utility management without sufficient consideration for human-centric design. Gaps exist in our understanding of how AI-driven interventions influence the daily lives and societal norms of urban dwellers. Investigating these behavioral outcomes can guide the creation of genuinely people-focused cities, where technology enhances rather than dictates the human experience.

Another significant knowledge gap is the adaptability of AI systems to diverse urban environments. Cities, characterized by varying cultural, economic, and regulatory landscapes, pose unique challenges to the universality of AI solutions. The assumption that a model developed in one locale can be directly applied to another overlooks the nuanced needs and constraints of different regions. What works for a technology-forward metropolis might not be suitable for a smaller, resource-constrained city struggling with basic infrastructure issues. Therefore, research must focus on developing adaptable AI frameworks that allow for localization and customization while maintaining core efficiencies.

The seamless integration of AI into existing urban infrastructures also presents knowledge gaps. Many cities grapple with legacy systems that may not be directly compatible with cutting-edge AI technologies.

Understanding how to retrofit these systems or develop cost-effective transition strategies remains a significant area needing further exploration. Additionally, there is a need for research into the lifecycle environmental impacts of AI deployments, from hardware manufacturing to electronic waste management, ensuring urban AI ecosystems meet sustainability goals.

Ethical concerns surrounding AI usage in urban settings add another layer of complexity, as they encompass privacy, autonomy, and equity issues. While ethical frameworks are being developed, the dynamic nature of AI technology means these guidelines can quickly become outdated or insufficient. Research needs to delve deeper into ethical considerations, particularly focusing on protecting vulnerable groups from potential biases and ensuring equitable access to AI-enhanced services. This involves interdisciplinary collaborations among technologists, policymakers, ethicists, and community leaders to create agile ethical norms that evolve alongside technological progress.

Knowledge gaps also persist in understanding the economic implications of AI in urban development. While there's considerable focus on job displacement, there's less clarity around the types of new jobs created by AI and how cities can prepare their workforce for these roles. Identifying pathways to reskill or upskill populations and exploring how AI can create economic opportunities in previously underserved sectors or communities is an area ripe for further investigation.

Finally, as cities worldwide implement AI strategies independently, there's a noticeable gap in shared knowledge dissemination and collaborative learning. Urban centers can greatly benefit from understanding heterogeneous approaches, highlighting successes and challenges faced by their counterparts globally. Encouraging and institutionalizing cross-city collaborations will be paramount for accelerating learning curves and avoiding redundant failures.

As we navigate these knowledge gaps, it's crucial to embrace a multidisciplinary approach drawing from technology, social sciences, humanities, and beyond. Addressing these knowledge deficiencies will empower cities to intelligently harness AI's transformative potential, ensuring not only technologically advanced but also socially responsible urban futures. Through targeted research efforts, we can build smarter cities that resonate with the people's realities residing within them while adapting quickly and efficiently to future challenges. By continuously updating our understanding and expanding the scope of our inquiries, we ensure that AI's role in urban development is guided by informed decisions, ethical considerations, and a vision that values inclusivity and sustainability.

Encouraging Continued Inquiry

In an era where AI is rapidly reshaping the urban landscape, the journey of discovery is only just beginning. It's essential for tech enthusiasts, urban planners, policymakers, and everyday citizens to keep questioning and exploring the myriad possibilities AI brings to our cities. By remaining curious and motivated to understand both the opportunities and challenges AI presents, we can create a vibrant culture of learning and adaptation. This ongoing inquiry will not only help us anticipate and manage the impacts of AI on urban life but also drive us towards smarter, more sustainable cities. Embracing a mindset of continued inquiry encourages collaboration across disciplines, propelling us toward innovative solutions that meet the evolving needs of urban environments while ensuring social equity and inclusiveness. Let this be a call to action for all those invested in the future of our cities, prompting a shared journey into the uncharted territories of urban AI development.

Collaborative Research Initiatives emerge as a cornerstone in the quest for advancing urban AI solutions, focusing the efforts of diverse stakeholders on common goals in reimagining city living. This kind of collaboration is not just beneficial; it's essential. When individual efforts converge through collaboration, they amplify their impact exponentially, creating momentum that's hard to replicate when working in isolation.

In an era where cross-discipline synergy is crucial, collaborative research initiatives provide a fertile ground for innovative solutions. Bringing together experts from AI, urban planning, sociology, environmental science, and government policy allows for a holistic approach to the smart cities movement. Each discipline offers its unique perspective, addressing specific aspects of urban challenges and ensuring that solutions are well-rounded and inclusive.

Consider the potential when academic institutions, government agencies, private corporations, and non-profits unite their resources and expertise to tackle urban challenges. These partnerships can lead to the development of cutting-edge technologies and the generation of robust data essential for testing new urban frameworks. Universities, with their academic rigor and research infrastructure, often take the lead, setting up think tanks and innovation hubs that prioritize city-centric AI advancements.

Notably, these collaborations thrive on mutual benefits and shared interests. For academic institutions, participating in such initiatives often leads to access to real-world data and practical challenges that enrich their research endeavors. Governments gain innovative solutions to policy-making while addressing urban inefficiencies, and tech companies can test and refine their emerging products, tailoring them for public sector implementation.

One tangible example of successful collaboration is the development of AI-driven traffic management systems. When cities work with tech firms and universities, they not only improve traffic flow but also gather invaluable insights into commuting patterns and urban dynamics. Such insights inform future urban planning and technology deployment, creating a virtuous cycle of improvement and adaptation.

Funding often poses a significant barrier to research, but collaborative initiatives open new avenues for investment. Public-private partnerships can attract substantial financial resources, with each partner contributing funds, expertise, or technology to a collective pool. In turn, this pooled investment attracts additional interest from venture capitalists and international funding bodies seeking to invest in promising and sustainable urban futures.

International partnerships further expand the possibilities. Sharing knowledge and resources across borders enables smart city innovations tailored to diverse contexts and challenges. Collaborative research initiatives can harness global perspectives, adapting them to local needs while ensuring that lessons learned in one part of the world are not lost but adapted and enhanced elsewhere. This global exchange accelerates the pace of innovation, ensuring that cities worldwide benefit from cutting-edge technologies.

Additionally, collaborative initiatives often drive the standardization of protocols and technologies, essential for interoperability and integration across urban systems. Shared research encourages the development of open-source platforms, reducing barriers to entry for innovators and ensuring that solutions are scalable and adaptable. As technological hurdles are lowered, it becomes easier for more cities, regardless of size or wealth, to participate in the smart city evolution.

Moreover, fostering a culture of collaboration encourages transparency, trust, and inclusivity in the research process. When diverse stakeholders unite, they embed local community needs and priorities into the research agenda. This inclusivity helps ensure that solutions are equitable and sensitive to socio-economic variations, placing people at the center of urban innovation. By incorporating feedback loops with community members, collaborative research ensures technologies serve the entire population and not just a privileged few.

However, these endeavors aren't without challenges. Synchronized collaboration across various entities requires alignment in vision, goals, and methodologies. Effective communication and goal-setting strategies are paramount. Regular meetings, clear frameworks for cooperation, and flexible structures allow partners to adjust strategies as projects evolve or as external factors change.

The momentum created by these initiatives not only helps solve current urban challenges but also lays down frameworks for future research and innovation. As cities continue to grow and evolve, so will the nature and complexity of the problems they face. Collaborative research, ever-evolving and adaptive, remains a dynamic force in navigating these changes, ensuring urban centers of tomorrow are as efficient as they are equitable.

Ultimately, the success of collaborative research initiatives in driving urban transformation lies in viewing cities not as fragmented entities but as interconnected systems. This perspective encourages researchers to think beyond isolated solutions and focus on systems thinking, recognizing the ripple effects of innovations across urban life. In doing so, cities become living laboratories, continuously experimenting, learning, and evolving in partnership with the very communities they serve.

Chapter 25: Global Implications of Urban AI Integration

As cities around the globe increasingly integrate artificial intelligence into their infrastructure, the global implications are both profound and varied. In developing nations, AI holds the promise of speeding up urban development and increasing access to essential services where traditional methods might falter. Urban planners and policymakers worldwide can gain valuable insights from each other's strategies, recognizing that equitable urban growth leveraged through AI demands cooperation and adaptability across borders. There's an immense opportunity to transform cities into hubs of innovation, economic prowess, and sustainability by sharing knowledge and methodologies. However, the road to integration is not without challenges. Global policymakers must address social equity, ensuring that advancements in AI don't widen existing disparities. Through thoughtful regulation and strategic international partnerships, urban AI deployments can lead to smarter, more inclusive cities that serve as a model for future urban development worldwide.

Impact on Developing Nations

As cities worldwide embrace AI integration in urban environments, developing nations stand at a crossroads. On one hand, AI presents an unprecedented opportunity to leapfrog traditional developmental hurdles. On the other, there are significant challenges that need addressing to ensure these technologies benefit all citizens equitably.

In developing countries, urban areas often grapple with rapid population growth, inadequate infrastructure, and limited resources. AI can help tackle these issues by optimizing existing systems and providing innovative solutions to urban challenges. For instance, intelligent algorithms can enhance traffic management, reducing congestion and pollution while improving public transportation efficiency. This not only makes commuting easier but also contributes to a better quality of life for city dwellers.

Healthcare systems in many developing nations can also witness transformational changes with AI. AI-powered diagnostics and telemedicine can bridge the gap in healthcare access, especially in remote and underserved areas. By enabling swift and accurate diagnosis, AI can alleviate the pressure on healthcare professionals and improve the quality of care patients receive. Such strides are crucial in locales where access to expert medical practitioners is often limited.

Furthermore, AI can revolutionize education by providing personalized learning experiences tailored to students' individual needs. In regions with overcrowded classrooms and scarce educational resources, AI-driven tools can facilitate more effective learning by adapting content to suit different learning paces and styles.

However, a successful AI integration requires not only technological innovation but also supportive policies and governance structures. Developing nations must invest in robust data infrastructure, ensure data privacy, and implement regulations to protect citizens' rights in the digital age. This necessitates international cooperation and knowledge transfer to establish frameworks that accommodate local contexts while adhering to global standards.

Economic considerations are pivotal. AI has the potential to drive economic growth by creating new industries and jobs. However, there is an accompanying risk of displacement for those employed in more traditional roles. To mitigate these effects, developing nations should prioritize reskilling and upskilling initiatives, ensuring the workforce can transition smoothly into emerging sectors.

The digital divide is another pressing issue. While AI can democratize access to information and services, it can also exacerbate existing inequalities if not implemented inclusively. Ensuring equitable access to AI technologies across different socioeconomic groups is essential, as reliance on AI without widespread access can lead to further social stratification.

AI in agriculture presents opportunities for rural development. By employing AI-driven analytics for weather prediction, crop monitoring, and supply chain management, developing nations can improve food security and agricultural productivity – a vital step toward economic resilience.

Despite the hurdles, the integration of AI in urban settings within developing nations offers a unique chance to address systemic issues that have persisted for decades. Governments, in collaboration with private sectors and international bodies, must work diligently to craft policies that foster innovation while ensuring citizen welfare.

An effective strategy lies in fostering collaborations between local tech innovators and global leaders in AI. Such partnerships can facilitate the exchange of best practices and emerging technologies while nurturing local talent through education and mentorship programs. Encouraging the local adaptation of AI solutions ensures that they are culturally relevant and address specific urban challenges faced by developing nations.

Ultimately, the impact of AI integration into urban environments in developing nations is profound. It is a transformative force poised to redefine the landscape of cities across these regions. However, the path forward requires deliberate planning, inclusive policies, and a commitment to equity and sustainability to ensure that the benefits of AI are accessible to all members of society.

Looking ahead, as AI continues its trajectory of change, developing nations have the opportunity to redefine their urban futures. By strategically leveraging AI, these nations can not only address current urban challenges but set the stage for sustainable, intelligent urban growth that empowers their citizens well into the future.

Lessons for Global Policymakers

As urban AI integration reshapes cities worldwide, global policymakers stand at a crossroads where strategic foresight can lead to transformative impacts. Policymakers need to strike a balance between leveraging AI for urban innovation and ensuring these advancements align with societal values and ethical considerations. It's crucial to establish robust international collaborations that foster shared best practices and standard regulatory frameworks, ensuring AI's benefits are accessible across diverse urban landscapes, from bustling metropolitan areas to emerging cities. Encouraging investment in AI literacy and infrastructure, while prioritizing transparency and inclusivity, will empower communities and safeguard citizens' trust. Global policymaking should aim for an adaptable and dynamic approach, focusing on long-term resilience and sustainability to amplify AI's potential in enhancing the quality of urban life for everyone. Embracing a proactive, inclusive strategy can unlock new opportunities, bridging gaps between technology and society, and paving the way for a smarter, more equitable urban future.

Promoting Equitable Urban Development is crucial in the context of urban AI integration. As cities around the world evolve into technological hubs, ensuring that development remains equitable is not simply a moral imperative—it's a necessity for sustainable progress. Policymakers must consider the implications of AI and technology in urban settings, acknowledging that the benefits can be unintentionally skewed towards more privileged populations. The gap between high-income and low-income areas can easily widen if technological advancements aren't implemented inclusively.

One of the core challenges is access to technology itself. In cities where resources are unevenly distributed, access to AI-driven solutions might be limited for lower-income communities. Such disparities can exacerbate existing inequalities, as these communities may not benefit from innovations that others have ready access to. AI solutions, which could transform sectors like transportation, healthcare, and education, might remain out of reach for those who need them most. Hence, urban planners and policymakers need to make concerted efforts to bridge these gaps.

Initiatives aimed at promoting equitable urban development should begin with community engagement. Residents from all socio-economic backgrounds must be involved in conversations about how AI could and should transform their living environments. Participatory planning ensures that the voices of all community members are heard and their needs are addressed. Policymakers might consider holding workshops, surveys, and forums to gather input from diverse populations. This grassroots involvement is vital in tailoring AI applications to serve the entire populace.

Education plays a pivotal role in promoting equitable urban development. Providing educational opportunities focused on digital literacy and AI can empower underprivileged communities, enabling them to participate actively in the digital economy. Schools and community centers could collaborate to offer courses and workshops that demystify technology and highlight the possibilities it can bring to everyday life. When community members understand AI technology, they are better positioned to leverage its benefits effectively.

Moreover, equitable infrastructure development cannot be overlooked. As urban spaces integrate AI, the physical layout and connectivity of city infrastructure need to be considered. Areas with limited internet connectivity or poor transport links might lag in adopting AI-enhanced services. Investing in improving infrastructure across all areas ensures that every citizen can access AI-driven solutions like smart healthcare services or intelligent public transport systems. Such investments can lead to overall improvements in quality of life and economic opportunity.

Financial mechanisms also play a critical role in fostering equitable urban development. Policymakers should explore funding models that prioritize projects with social equity impacts. For example, public funding could be allocated to initiatives aimed at installing AI technologies in underserved areas, such as deploying AI-driven environmental monitoring systems in neighborhoods suffering from pollution. Additionally, incentivizing private sector partnerships with clear equity objectives can attract investments that align with social goals.

Global policymakers stand to gain essential insights from observing successful localized initiatives. Cities worldwide can serve as examples or cautionary tales depending on how they manage urban AI integration. For instance, examining Scandinavian smart city models could provide lessons on implementing welfare-centric approaches that emphasize sustainability and equality. Urban policymakers can adapt these lessons to their unique social and economic environments, focusing on crafting adaptable strategies that reflect local dynamics.

Furthermore, international cooperation is paramount in learning and sharing best practices for equitable urban AI development. Nations can collaborate through platforms, share data, and support research that underscores the importance of balanced urban growth. Such alliances might lead to the establishment of international guidelines or standards for equitable development in AI integration. By collectively acknowledging the global nature of these challenges and working towards inclusive solutions, cities can create technologies that serve everyone fairly.

Ultimately, promoting equitable urban development means recognizing the interconnectedness of technology and society. As policymakers harness the power of AI, their strategies should prioritize closing socio-economic divides while leveraging technology's transformative potential to elevate the quality of urban life for all. Through deliberate and inclusive actions, we can shape cities that are not only smart but also just, opening doors for every citizen to enjoy the benefits of an AI-infused urban landscape.

Conclusion

As we stand at the crossroads of technological and urban evolution, the narrative of cities is being rewritten. Artificial Intelligence has become a transformative force reshaping the contours of urban life and offering pathways to smarter, more responsive environments. It's clear that AI holds immense promise—revolutionizing sectors ranging from traffic management to public safety, healthcare, and beyond.

In this journey through the myriad facets of AI in urban settings, our exploration reveals a tapestry of innovation interwoven with profound challenges. On one hand, AI continues to enhance efficiency and sustainability in cities. Intelligent systems optimize energy consumption, provide robust public safety solutions, and enable predictive urban planning that anticipates growth and change with remarkable precision.

Yet, these advancements come with their share of hurdles. Ethical considerations loom large, especially concerning data privacy and surveillance. The balance between harnessing data for good and protecting individual rights is delicate and paramount. Policymakers and urban planners must navigate these murky waters to ensure AI is implemented in a way that is equitable and inclusive.

The future of urban life, underpinned by AI, is one of interconnected systems and substantial promises. Smart grids are reducing energy waste, and AI-driven emergency response systems are improving safety and security. The integration of AI in healthcare offers unprecedented access to services, bringing innovative solutions for remote diagnostics and personalized patient care directly into the urban framework.

Moreover, AI's role in digital governance and citizen engagement is pivotal. By facilitating transparent governance and enhancing public services, AI invites more active civic participation. It empowers citizens, making them pivotal players in the constant remaking of their cities and encouraging a shared responsibility for urban development.

At the heart of these innovations lies the need for open collaboration and a multidisciplinary approach. Effective AI integration in smart cities requires robust partnerships across the public and private sectors, as well as international cooperations. This collective effort is vital for overcoming barriers and maximizing the potential benefits of AI technologies.

One can't overlook the profound economic impacts brought on by AI. While automation and AI-driven processes may disrupt traditional job sectors, they also foster new opportunities, driving economic development and innovation. As cities adapt, they serve as fertile grounds for startups and emerging tech enterprises, exemplifying resilience and adaptability in the face of changing economic landscapes.

As we envision a future where cities are not just smart but also sustainable and inclusive, we must embrace both the opportunities and responsibilities that come with AI. It's imperative that urban development be guided by ethical frameworks and regulatory standards that foster innovation while safeguarding communities.

Looking ahead, the global implications of urban AI integration signal a shift in how cities around the world approach development. In developing nations, AI presents an opportunity to leapfrog into more advanced systems, promoting equitable growth. For global policymakers, these trends offer invaluable lessons in scalable and sustainable urban planning.

Ultimately, the story of AI in urban spaces is not confined to technological advancements alone. It's a story of evolving human experiences within the urban fabric—a dynamic blend of tradition and innovation. As AI redefines the way we live and interact with our cities, it also invites us to reconsider notions of community, identity, and belonging. The challenges are significant, but so too are the possibilities for creating urban environments that echo the aspirations and dreams of their inhabitants.

Our exploration concludes with a hopeful vision: a future where cities harness the power of AI not just to solve problems but to enhance the very quality of urban life. As we embrace this transformation, the hope is that our cities will become more vibrant, inclusive, and resilient—adapting not just to the needs of today but anticipating those of tomorrow.

Appendix A: Appendix

The appendix serves as a comprehensive resource for readers, providing additional materials that complement the main text. It includes detailed charts, data tables, and figures that can enhance understanding of the complex dynamics discussed in the book. Through these resources, readers can gain deeper insights into how AI technologies are reshaping urban environments. Additionally, supplementary notes and references are included to assist in exploring key topics further. This section is a valuable tool for urban planners, policymakers, and tech enthusiasts seeking to delve into the specifics and nuances of AI's role in creating smarter and more sustainable cities. By offering a mix of visual aids and explanatory notes, the appendix enriches the reader's journey through the myriad ways in which AI is innovating and transforming urban life.

Glossary of Terms

This glossary provides definitions and explanations of key terms related to the use of artificial intelligence and emerging technologies in urban transformations. Understanding these terms will help clarify how AI is shaping smarter, more sustainable cities.

- **Artificial Intelligence (AI):** A field of computer science focused on creating systems capable of performing tasks that typically require human intelligence, such as decision-making, visual perception, and language understanding.

- **Smart City:** An urban area that uses various types of electronic methods, including sensors and data analysis, to efficiently manage resources and services, often aiming to improve the quality of life for its citizens.
- **Internet of Things (IoT):** A network of physical devices, vehicles, buildings, and other items embedded with electronics, software, sensors, and connectivity, enabling them to collect and exchange data.
- **Machine Learning:** A subset of AI that involves the development of algorithms allowing computers to learn from and make decisions based on data.
- **Predictive Policing:** The application of analytical techniques to identify potential criminal activity, allowing law enforcement to allocate resources effectively.
- **Smart Grid:** An electricity network enabling a two-way flow of electricity and data with digital communications technology, aimed at improving reliability, efficiency, and sustainability of energy distribution.
- **Autonomous Vehicles:** Self-driving cars and other vehicles that use AI to navigate and operate without direct human control.
- **Urban Mobility:** Systems and services that facilitate movement within urban areas, often optimized through AI technologies to reduce congestion and improve transportation efficiency.
- **Digital Governance:** The application of digital technologies to improve the processes and services delivered by government

to citizens, ensuring transparency and enhancing public engagement.

- **Data Privacy:** The protection of personal information stored digitally, especially in the context of smart city technologies that collect vast amounts of citizen data.
- **Sustainable Energy Solutions:** Methods and technologies, often powered by AI, aimed at reducing environmental impact and increasing the efficiency of energy usage in urban areas.
- **Urban Planning:** A technical and political process focused on the development and design of land use in an urban environment, increasingly employing AI for predictive modeling and infrastructure development.
- **AI Ethics:** The branch of ethics concerned with how AI technologies are developed and deployed, ensuring that they are used responsibly and do not infringe on human rights.
- **Public-Private Partnerships:** Collaborations between government entities and private sector companies to finance, build, and operate projects, particularly in the context of developing smart city initiatives.
- **Citizen Engagement:** The process of involving citizens in decision-making, planning, and development processes in urban environments, often facilitated by digital platforms.
- **Resilience:** The ability of urban systems to adapt to and recover from challenges such as natural disasters, often bolstered by AI technologies that enhance preparedness and response.
- **Economic Transformation:** The changes in economic structures and practices occurring as a result of AI

implementation in urban areas, often leading to new job creation and shifts in traditional roles.

Resources for Further Reading

Diving deeper into the fascinating intersection of artificial intelligence and urban development opens a vast world of knowledge and literature. Whether you're a tech enthusiast, an urban planner, or just someone curious about how AI is reshaping city life, there's plenty of additional reading to enrich your understanding. Here are some resources that delve into key concepts, offer theoretical insights, and showcase practical applications.

Starting with foundational knowledge, books like *Artificial Intelligence: A Guide to Intelligent Systems* by Michael Negnevitsky provide an excellent overview of AI principles and their broad scope of application. You might also consider Stuart Russell and Peter Norvig's *Artificial Intelligence: A Modern Approach*, a seminal text that thoroughly covers AI's essential algorithms and techniques. These resources are particularly useful for understanding how AI underpins many smart city technologies.

For those interested in the real-world application of AI in smart cities, the book *The Smart Enough City: Putting Technology in Its Place to Reclaim Our Urban Future* by Ben Green is a thought-provoking read. Green argues for a balanced approach in integrating technology within urban settings, stressing the importance of human-centric decision-making. This book emphasizes that while technology can enhance urban living, it must be guided by ethical considerations and community involvement.

Looking at smart transportation systems, *Transport, Transformed: AI in Urban Mobility Systems*, edited by Jonathan Tennyson, explores various AI solutions being deployed to tackle urban transport challenges. This anthology includes case studies from cities around the world, offering both successes and areas for improvement.

To understand more about AI's role in sustainable urban development, delve into *Smart Green Cities: Using Urban Analytics to Combat Climate Change* by Woodrow Clark and Grant Cooke. The book highlights how urban analytics and AI innovations can vitalize efforts toward achieving sustainable urban environments. It stresses data-driven strategies for managing energy resources and improving city infrastructure.

Datasets form the backbone of AI systems, and understanding how they're managed is crucial. Read *The Data-Driven City: Big Data and Urban Analytics* by Rob Kitchin to comprehend how cities collect, manage, and utilize data in innovative ways. It provides insight into the data governance frameworks necessary for ensuring privacy and security within smart cities.

On a more technical front, *Deep Learning* by Ian Goodfellow, Yoshua Bengio, and Aaron Courville is an essential text for those who wish to understand the algorithms powering AI systems. While the focus is broad, its applications in predictive modeling, one of the vital tools in urban planning and resource management, can be particularly beneficial.

Consider *Ethical AI: Creating a Framework That Balances the Benefits of Machine Learning with the Risks* by James Loudermilk for a comprehensive look at the ethical challenges involved in deploying AI in urban settings. The text offers frameworks for ethical decision-making and addresses concerns related to surveillance, privacy, and algorithmic bias.

In the realm of public policy, *Smart Cities: Big Data, Civic Hackers, and the Quest for a New Utopia* by Anthony M. Townsend explores the intersection of technology, governance, and citizen engagement. It's an inspiring narrative that discusses the social and political dimensions of smart city initiatives, offering lessons for policymakers and civic leaders alike.

For insights into the future directions of AI in urban life, *The Rise of the Platform City: Urban Algorithms, Mass Participation, and the Future of Infrastructure* by Sam Lowry provides a speculative look at future urban landscapes. It postulates how platform-based governance could revolutionize infrastructure management and public engagement.

Lastly, engaging with online platforms like IEEE Xplore and the Journal of Urban Technology can be invaluable for up-to-date research articles on the latest technological advancements and case studies in urban AI applications. These academic journals frequently cover emerging trends and provide evidence-based insights into smart city technologies.

This list is by no means exhaustive, but these resources should offer a solid starting point for delving deeper into the complex, evolving domain of AI within urban contexts. As you explore these materials, you'll gain a broader perspective on how AI is contributing to building smarter, more resilient cities and the profound impact it has on our current and future urban experiences.

www.ingramcontent.com/pod-product-compliance
Lightning Source LLC
Chambersburg PA
CBHW051237050326
40689CB00007B/948

* 9 7 8 1 4 5 6 6 5 9 7 6 9 *